T0302231

Understanding Risk to Wildlife from Exposures to Per- and Polyfluorinated Alkyl Substances (PFAS)

Understanding Risk to Wildlife from Exposures to Per- and Polyfluorinated Alkyl Substances (PFAS)

Mark S. Johnson, Michael J. Quinn Jr., Marc A. Williams, and Allison M. Narizzano

US Army Public Health Center, Toxicology Directorate, Aberdeen Proving Ground, MD, USA

CRC Press
Taylor & Francis Group
Boca Raton London New York

CRC Press is an imprint of the
Taylor & Francis Group, an **informa** business

First edition published 2021 by
CRC Press
6000 Broken Sound Parkway NW, Suite 300, Boca Raton, FL 33487-2742

and by CRC Press
2 Park Square, Milton Park, Abingdon, Oxon, OX14 4RN

Library of Congress Cataloging-in-Publication Data

Names: Johnson, Mark S. (Toxicologist), author. | Quinn, Michael J., Jr.,
author. | Williams, Marc A. (Biologist), author. | Narizzano, Allison
M., author.
Title: Understanding risk to wildlife from exposures to per- and
polyfluorinated alkyl substances (PFAS) / Mark S. Johnson, Michael J.
Quinn Jr., Marc A. Williams, Allison M. Narizzano.
Description: First edition. | Boca Raton : CRC Press, 2021. | Includes
bibliographical references and index.
Identifiers: LCCN 2020049028 (print) | LCCN 2020049029 (ebook) | ISBN
9780367754075 (hardback) | ISBN 9781003162476 (ebook)
Subjects: LCSH: Perfluorinated chemicals--Environmental aspects. |
Perfluorinated chemicals--Bioaccumulation. | Perfluorinated
chemicals--Toxicology. | Animals--Effect of chemicals on. |
Animals--Effect of pollution on. | Environmental toxicology. |
Ecological risk assessment.
Classification: LCC QH545.P38 J64 2021 (print) | LCC QH545.P38 (ebook) |
DDC 363.738/4--dc23
LC record available at https://lccn.loc.gov/2020049028
LC ebook record available at https://lccn.loc.gov/2020049029

ISBN: 978-0-367-75407-5 (hbk)
ISBN: 978-0-367-75427-3 (pbk)
ISBN: 978-1-003-16247-6 (ebk)

Typeset in Times
by Deanta Global Publishing Services, Chennai, India

Contents

Foreword

This publication was prepared for educational use. The views and assertions expressed in this presentation are the private views of the authors and do not necessarily reflect, nor should they be construed as reflecting the views, and official policy of the Department of Defense, the Department of the Army, the US Army Medical Department or the US Federal Government. Furthermore, use of trademarked name(s) does not imply endorsement by the US Army but is intended only to assist in identification of a specific product.

Preface

Society has been concerned with unintended effects of anthropogenic chemicals on wildlife species since Rachel Carson's *Silent Spring*, associated with the environmental release of certain industrial compounds and pesticides. Many regulations now exist to protect wildlife from exposure-associated adverse effects; however, assessing the potential for toxic effects to wildlife is not straightforward. Often, an incomplete understanding of the factors that are important in estimating exposure to the available toxicity data are limited to a few species within a class of vertebrate animals, which makes assumptions critical in estimating risk.

Many biological pathways are evolutionarily conserved among species, making some assumptions useful, yet physiological differences between species can add to difficulties in extrapolating toxicity data. In this book, we have focused on a set of chemicals of emerging concern, that are collectively termed per- and polyfluorinated alkyl substances (PFAS). Due in large part to the recalcitrant nature of some PFAS, they have been found in the tissues of wildlife species the world over. This observation makes it even more important to recognize, prioritize, and mitigate environmental sources of concern through the risk assessment process.

This book attempts to provide an in-depth review of PFAS toxicology relevant to terrestrial vertebrates, and derives toxicity reference values (TRVs) to assist this process. Two levels of TRVs (high and low) are provided for both screening and baseline level assessments. Together, these TRVs and their associated supporting information are intended to support investigations of contaminated sites and to assist with the prioritization and mitigation of potential sources of environmental contamination.

Acknowledgments

The work provided in this book was supported by the US Army Environmental Command, Installation Restoration Program, and the Strategic Environmental Research and Development Program (SERDP), Project No. E-2625 (Development of Toxicity Data to Support Mammalian Toxicity Reference Values for Perfluorinated Compounds).

Authors

Mark S. Johnson, Ph.D., DABT, Fellow ATS
Director, Toxicology, US Army Public Health Center, Aberdeen Proving Ground, MD, USA

Dr. Johnson currently serves as the Director of Toxicology, US Army Public Health Center at Aberdeen Proving Ground, MD, USA, where he is responsible for the operational and technical arm of the Army Surgeon General and the Assistant Secretary of the Army for toxicological matters. He has worked extensively in the evaluation of the toxicity of unique military compounds and the development and evaluation of a phased approach to the gathering of toxicity data for new compounds under development. He has authored over 100 peer-reviewed publications, book chapters, and technical reports, and serves on several NATO and EPA panels. He has been a member of Society of Environmental Toxicology and Chemistry (SETAC) since 1997 and is a past Steering Group Member of the Wildlife Toxicology World Interest Group, past chair of Ecological Risk Assessment World Interest Group, and a member of the World Science Committee for SETAC and SETAC North America. Dr. Johnson is also a member of the International Board of Environmental Risk Assessors (IBERA). He has been a member of the Society of Toxicology since 2009.

Dr. Johnson is a fellow of the Academy of Toxicological Sciences, Chair of the Tri-Service Toxicology Consortium (TSTC), past Steering Committee Chair of the Joint Army-Navy-NASA-Air Force (JANNAF) Propulsion Committee, Subcommittee on Safety and Environmental Protection, and a past president of the American Board of Toxicology (ABT).

Michael Quinn, Jr., Ph.D.
Chief, Health Effects Division, Toxicology Directorate, US Army Public Health Center, Aberdeen Proving Ground, MD, USA

Dr. Michael Quinn, Jr., is the US Army Public Health Center's Health Effects Division Chief for the Directorate of Toxicology. He has contributed to numerous toxicology studies at the APHC on explosives and propellants on a wide range of taxa, that has included mammals, birds, reptiles, and amphibians. His areas of expertise are developmental and reproductive toxicology and endocrine disruption. Dr. Quinn's primary passion is method development, and he has contributed to developing an avian two-generation toxicity test for the US Environmental Agency's Endocrine Disruptor Screening Program's Tier 2 battery of tests. Currently, he is leading a series of toxicity assessments for per- and polyfluoroalkyl substances (PFAS), associated with contamination from aqueous film-forming foam (AFFF) use, and for PFAS-free AFFF replacements.

Allison M. Narizzano, Ph.D.
Biologist, Toxicology, US Army Public Health Center, Aberdeen Proving Ground, MD, USA

Dr. Narizzano is a biologist at the US Army Public Health Center at Aberdeen Proving Ground, MD. She completed her Ph.D. in Toxicology at the University of

Maryland in 2020. In her doctoral work, Dr. Narizzano evaluated the effects of
defense-relevant chemicals, including PFAS and insensitive munitions, on non-tradi-
tional laboratory models.

Marc A. Williams, Ph.D., Fellow AAAAI
Biologist, Directorate of Toxicology, US Army Public Health Center, Aberdeen Proving Ground, MD, USA

Dr. Williams is a biologist (immunotoxicologist) and serves as a Project Manager
for the Wildlife Toxicity Assessment Program, Toxicology Directorate – Health Effects
Division, where he provides leadership and subject matter expertise in applied toxicology and environmental toxicity assessments of unique military chemicals of interest,
that include PFAS, nanomaterials, particulate matter, and complex chemical mixtures.
Dr. Williams has previously served as a research biologist at the US Environmental
Protection Agency (US EPA), with research interests in human health and toxicological exposures impacting wellness and protective immune competence. Prior to joining
the US EPA, Dr Williams was an Assistant Professor of Medicine and Environmental
Medicine at the University of Rochester School of Medicine and Dentistry, New York,
and an Instructor of Medicine at the Johns Hopkins University School of Medicine,
where he led research programs in human immunology, applied toxicology, and
models of human disease. Dr. Williams holds primary degrees, firstly in Molecular
Biotechnology and Process Bioengineering, and secondly in Molecular Cell Biology
and Immunology. Dr. Williams earned his Ph.D. in Haematological Oncology and
Cancer Immunology from Queen Mary College, University of London, UK.

Dr. Williams is a Fellow of the American Academy of Allergy, Asthma and
Immunology (AAAAI) and served on their Special Emphasis Working Group to
highlight the importance of nanoparticle toxicology in human disease pathways. Dr
Williams has authored over 100 peer-reviewed articles, review papers, book chap-
ters, book titles, and formal scientific reports. Dr. Williams serves on several *ad hoc*
grant review panels for NASA aligned to their Atmospheric, Earth and Environmental
Health Research Panels, and currently serves as the Non-VA Federal Government
Chair of the Intergenerational Health Effects of Military Exposures Work Group. Dr.
Williams is a member of the Society of Environmental Toxicology and Chemistry
(SETAC) North America and the Society of Toxicology (SOT). He actively serves on
the Membership Committee (MC) of SETAC North America (SNA) and as MC liaison
to the Development Committee of SNA. For the past six years, Dr Williams has served
a leading role on the National Cancer Institute/NIH Nanomaterials Working Group,
and, for the past three years, as the Editor-in-Chief of *Drug and Chemical Toxicology*
and, for 10 years, as a Regular Editorial Advisory Board Member of *Toxicology Letters*.

1 Introduction

Mark S. Johnson

CONTENTS

Predicting the bioaccumulation of PFAS in terrestrial wildlife (including humans) is proving to be extremely complex. As a group, PFAS act differently from traditional non-ionic organic molecules, where PFAS can breakdown and reform, while some have been demonstrated to be extremely persistent. Toxicokinetic profiles for many PFAS vary widely between species and sexes (Li *et al.* 2017, Conder *et al.* 2008). In some cases, their structure resembles that of fatty acids, and they tend to accumulate in the liver rather than partitioning to fats. Given the lack of understanding of how some lipids function on a molecular level, the mechanisms of toxicity of the PFAS effect are poorly understood.

The extent of bioaccumulation may be the best predictor of toxicity. As exposure is sustained, so is the systemic dose to the target site. Largely, females of mammalian species may be more effective at excretion than males, given the larger number of pathways for release (e.g., menstruation, parturition, and lactation). Some species of phylogenetically similar taxa can show marked differences in bioaccumulation (or biological half-lives) and mechanisms for these differences between species are also poorly understood.

PFAS have been found in various tissues of many species of wildlife, including ring-billed gulls (*Larus delawarensis*), California sea lion (*Zalophus californianus*), polar bear (*Ursus maritimus*), Laysan albatross (*Phoebastria immutabilis*), bottlenose dolphin (*Tursiops truncatus*), and various fish species, samples of which reveal the global extent of PFAS transport (Giesy and Kannan 2001, Kannan *et al.* 2002a,b). Thus, PFAS are globally distributed, and are detected in water, air, house dust, soil, sediment, sludge from wastewater treatment plants, and biosolids, and are largely non-degradable.

Many of the more persistent PFAS are no longer being produced by industry, which places greater importance on accurately characterizing the risk associated with potential environmental sources of previous releases. Many jurisdictions have regulations that require the use of risk assessments to determine whether contaminated media should be remediated. Traditionally, exposures to important wildlife species are modeled, or, less often, concentrations are measured and compared with a toxicity benchmark to determine the degree of hazard or risk. Here, we provide the extent of knowledge regarding effects to terrestrial wildlife, defined here as terrestrial mammals, birds, reptiles, and amphibians. Where sufficient data are provided, toxicity reference values (TRVs) are derived to assist in characterizing environmental sources and in making risk-based decisions.

DERIVATION OF TOXICITY REFERENCE VALUES (TRVS)

Derivation of TRVs requires a standardized approach to reviewing pertinent information and using this information to develop a value useful for screening and predicting risk. However, very few accounts published in the scientific literature have been developed with the goal of deriving TRVs. Therefore, clear interpretation of the results, the quality of the methods, and professional judgment regarding the potential adverse effect are key components to translate data to the risk to wildlife species and to derive useful TRVs. Largely, we begin with a clear approach to database query, where dates, databases searched, keywords used, number of hits, number of titles reviewed, number of abstracts retrieved, and publications reviewed are documented.

Controlled laboratory studies, where dose-response relationships are presented, are highly valued; however, other field data and mechanistic information, where biological pathways are conserved, can be used to help corroborate TRV estimates for interspecies extrapolation. When possible, benchmark dose methods are used to fit the data, and points of departures are determined typically at the sub-lethal benchmark dose, with 90% lower confidence levels ($BMDL_{10}$) for screening values, and the 50% confidence benchmark dose (BMD) level for baseline risk assessments, where decisions are made. Where models do not fit the dose-response data, the no-observed-adverse-effect and lowest-observed-adverse-effect levels, NOAEL and LOAEL, respectively, are used. Uncertainty factors are used according to those described in USACHPPM (2000), when data are incomplete for extrapolation. Hence, a TRV-low and a TRV-high are developed for screening and decision making, respectively. Only data within classes are used (e.g., mammal data for mammals) and only end points that are clearly considered to be adverse are included. Since any adverse effect has the potential to indirectly affect fitness (i.e., by influencing population regulation), we make no judgements (or adjustments) prioritizing toxicological end points where considered to be clearly adverse. Here, we follow the methods found in USACHPPM (2000) and others (Johnson and McAtee 2015, Deck and Johnson 2015).

Taxonomic diversity in toxicity data between species/genera/families and orders provides confidence in the utility of class-specific TRVs, particularly when effects and exposures are relatively consistent. Differences in methods can influence variation in the results and, as such, can be difficult to tease out from species differences. Therefore, each value also has a general level of confidence associated with it, to provide context to the user.

A particular challenge for this PFAS family of chemicals is that the toxicity is dependent, in many ways, upon the kinetics of exposure. Significant differences, particularly in excretion rates, exist between species and genders, and may also exist between age classes. Toxicity is dependent upon internal bioaccumulation of PFAS, that may vary between species, for many of which there are no empirical data. Some physiologically based pharmacokinetic models exist for rodents, primates, and humans for PFOA and PFOS (Loccisano et al. 2013, Chou and Lin 2019). Few other models exist for other species and PFAS, making extrapolation between species difficult.

TRVs are to be used as screening-level benchmarks for wildlife at or in close proximity to contaminated sites. The protocol for performing this assessment are described, in part, in Technical Guide No. 254 – the *Standard Practice for Wildlife Toxicity Reference Values* (USACHPPM 2000, Johnson and McAtee 2015, Deck and Johnson 2015).

2 Perfluorooctanoic acid (PFOA)

Marc A. Williams

CONTENTS

PERFLUOROOCTANOIC ACID AND ITS USES

A wide range of manufactured per- and polyfluoroalkyl substances (PFAS), including perfluorooctanoic acid (PFOA), have found broad utility in an array of consumer goods and industrial products, and are widely recognized as emerging pollutants, with heightened concern at their diverse toxicological and environmental impacts (Buck *et al.* 2011, Vierke *et al.* 2012, USEPA 2014, NTN 2015). The extensive use of perfluorinated compounds (PFCs) in surface coatings and protectant formulations since the early 1950s was due to their unique chemical properties (Buck *et al.* 2011).

TOXICOLOGICAL EFFECTS OF PFOA ON WILDLIFE

This chapter summarizes the toxicological effects of PFOA on wildlife following exposure to this compound and evaluates the PFOA toxicity data that are used to derive toxicity reference values (TRVs). The TRVs are to be used as screening-level benchmarks for wildlife at or in close proximity to contaminated sites. The protocol for performing this assessment is, in part, described in Technical Guide No. 254 – the *Standard Practice for Wildlife Toxicity Reference Values* (USACHPPM 2000, Johnson and McAtee 2015, Deck and Johnson 2015).

ENVIRONMENTAL FATE AND TRANSPORT

PFOA is manufactured as a fully fluorinated organic synthetic acid, that is used to synthesize fluoropolymers. PFOA is industrially synthesized by the Simons electrochemical fluorination (ECF) process or by telomerization (for details, please refer to Beesoon *et al.* (2011) and HSDB (2016)). Thus, all PFAS, including PFOA, which are found in the environment are anthropogenic substances and do not occur naturally. PFOA is moderately soluble in water, and both PFOA and its salts are considered to be insoluble in non-polar solvents. In addition, the water solubility of PFOA is enhanced when other ions are present (Ylinen *et al.* 1990, Ehresman *et al.* 2007).

The relatively weak acid dissociation constant (pK_a) for PFOA is log 2.8, and, at environmental pH values, PFOA is found in its dissociated form, perfluorooctanoate, which is a negatively charged ion. Additionally, unlike most other persistent organic pollutants that bioaccumulate, PFOA is moderately water-soluble and does not bind strongly to soil or sediments (Armitage *et al.* 2006). Thus, PFOA and other perfluoroalkyls tend to be highly mobile in soil and leach into groundwater and surface water columns, with broad potential for contamination of the local environment (ATSDR 2009). One of the greatest environmental concerns, with regard to the use of PFOA and other PFASs, is in the context of fire-fighting aqueous film-forming foam (AFFF) to suppress fires, that might be encountered on air force and army airfield installations (Darlington *et al.* 2018). The concern here is that deployment of AFFF concentrates on runways and airfield surfaces would leach chemicals into the immediate and local environments, posing significant health hazards to human health and wildlife; many of these installations or airfields are often located in highly populated areas and/or locations rich in wildlife diversity (Anderson *et al.* 2016, Larson *et al.* 2018).

The chemical characteristics of PFOA, with respect to experimental hydrolysis studies at pH 5, 7, or 9, indicate a half-life of 92 years (OECD 2002, 2006). Based on the data for PFOA, it is not expected that hydrolysis is an important degradation process for perfluorinated carboxylates and sulfonates in the environment. In addition, the available data indicate that perfluoroalkyl compounds are resistant to aerobic biodegradation. For example, PFOA was not biodegraded during an OECD guideline manometric respirometry-screening assay for ready biodegradability (Stasinakis *et al.* 2008). Others have reported that PFOA was not degraded at an initial concentration of 5 mg/L in aerobic sewage sludge in a laboratory-scale reactor (Meesters and Schröder 2004).

Although perfluoroalkyl acids are of relatively low volatility, their concentrations have previously been measured in outdoor air at a few locations in the United States, Europe, Japan, and over the Atlantic Ocean (Harada *et al.* 2005, 2006, Barton *et al.* 2006, Barber *et al.* 2007, Kim and Kannan 2007). Mean PFOA concentrations of between 1.54 and 15.2 pg/m^3 in air samples collected in urban locations of Fukuchiyama and Morioka, Japan, and in rural locations of Kjeller (Norway), Albany (New York), and Mace Head (Ireland). Higher mean concentrations (101–552 pg/m^3) were measured at urban locations in Oyamazaki, Japan and at urban locations in Manchester and semi-rural locations in Hazelrigg, United Kingdom. Maximum reported concentrations at Oyamazaki and Hazelrigg were 919 and 828 pg/m^3, respectively. The elevated concentrations at Hazelrigg were, in part, explained by emissions from a fluoropolymer production plant that was located 20 km upwind of this semi-rural community.

PFOA concentrations exceeded the quantitation limits for the chosen method (70,000–170,000 pg/m^3) in 6 out of 28 air samples collected along the fence line of the fluoropolymer manufacturing facility at DuPont Washington Works (near Parkersburg, West Virginia, in the Ohio River valley) (Barton *et al.* 2006). The reported concentrations in these six samples ranged from 75,000 to 900,000 pg/m^3. The highest concentrations were measured at locations downwind of the facility. High-volume air samples collected at several monitoring stations near the Washington Works facility contained PFOA at concentrations that ranged from 10 to 75,900 pg/m^3 (Barton et al., 2006). The mean and median of these reported concentrations were 5,500 and 240 pg/m^3, respectively. Nonetheless, despite the above data, Inhalation exposures are likely to be significant only for wildlife in close proximity to manufacturing locations.

Due in large part to the almost ubiquitous presence and prolonged environmental persistence of PFOA, it has been shown that PFOA is commonly found in wildlife, including all major organs, tissues, and bodily fluids, and, with the exception of gastric secretions, PFOA is also present in most biological fluids, in the form of the perfluorooctanoate anion – an important characteristic of PFOA that dictates its absorption and membrane transport (Buck *et al.* 2011, Butt *et al.* 2010, Ahrens 2011, Sturm and Ahrens 2010, Houde *et al.* 2006a, Butenhoff *et al.* 2006, Calafat *et al.*2007, Giesy and Kannan 2001, Kannan *et al.* 2004). Furthermore, the highest concentrations of PFOA were found in serum rather than in fat (Ehresman *et al.* 2007, Ylinen *et al.* 1990).

BIOACCUMULATION AND ELIMINATION

The bioaccumulation potential of PFASs is thought to increase with increasing chain length. In animals, perfluoroalkyls bind to the protein albumin in blood, liver, and eggs, and do not accumulate in fat tissue. However, PFOA has a very low serological elimination rate, with an estimated elimination half-life of approximately 75 days (ATSDR, 2009). The highest levels of perfluoroalkyls in animals have been measured in apex predators, such as polar bears, which indicates that these substances biomagnify in food webs. It is also recognized that, due to the persistence and capacity for long-term accumulation of PFOA, higher trophic level wildlife like fish, piscivorous birds, and Arctic biota are at risk of continuous exposure to PFOA (USEPA 2006, UNEP 2006).

PFOA was found at quantifiable levels in a range of wildlife and their natural habitats. The concentrations of PFOA that were measured in plasma were highest in juvenile bottlenose dolphins (*Tursiops truncatus*) sampled from two US southeast Atlantic sites (namely Charleston, SC and Indian River Lagoon, FL), with adult males containing the next-highest concentrations, and adult females showing the lowest concentrations of PFOA in the plasma (Fair *et al.* 2012). Female PFOA loads were likely lower because of their ability to offload organic pollutants to their young, while concentrations of PFOA in the plasma decrease with age. The source of PFOA and other PFASs found in young dolphins appears to be their mother's milk and their diet (Fair *et al.* 2012); elsewhere along the East Coast of the US, PFOA was detected in blood samples of all dolphins tested.

When dosed with up to 30 mg/kg of PFOA by oral gavage for 28 days, female rats eliminated nearly all the PFOA in urine during the first week of the study, but males did not. By the end of the study, male and female rats had eliminated almost all of the dosed PFOA. At the end of the study, males had higher concentrations of serum PFOA in all dosage groups, when compared with female rats (Hanhijärvi *et al.* 1987). In a further study, steers dosed once with 1 mg/kg of PFOA eliminated the entire dose in nine days. The primary elimination route was *via* the urine, with less than 5% eliminated *via* feces. Metabolites were not detected in plasma, urine, or feces (Lupton *et al.* 2012).

Studies of different animals showed that PFOA, in the form of ammonium perfluorooctanoate (PFOA), is well absorbed following oral or inhalation exposure, and, to a lesser extent, following dermal exposure (USEPA 2002). Male rats were fed PFOA for 13 weeks in a study where exposures were equivalent to 0.06, 0.64, 1.94, or 6.5 mg/kg. Serum PFOA levels increased proportionally with dietary exposure concentration; however, there was no increase in serum PFOA over the course of exposure (Perkins *et al.* 2004). Following four days of dermal exposure to PFOA in acetone, that was applied to the ears of mice, serum concentrations indicated that

PFOA penetrated the skin. Approximately 39% of the applied dose penetrated the skin, with 11% remaining within the skin tissue and 37% remaining at the skin surface (Franko *et al.* 2012).

In another relevant study, pregnant rats dosed orally with PFOA passed the PFOA along to their offspring. At the steady-state concentration, mean maternal serum concentrations were 11.2, 26.8, and 66.6 µg/mL, following oral exposure dose levels of 3, 10, and 30 mg/kg-d. Additionally, mean concentrations of PFOA in fetal plasma was collected on day 21 of gestation were approximately half the maternal mean serum concentrations, at 5.9, 14.5, and 33.1 µg/mL from dams with oral exposure dose levels of 3, 10, and 30 mg/kg-d, respectively (Hinderliter *et al.* 2005).

Although considerable information aligned with human health effects and the toxicological profile of PFOA is available from several documents published by the USEPA and in other published peer-reviewed articles (e.g., ATSDR (2009, 2015) and USEPA technical reviews), those will not be discussed in detail in the current wildlife toxicity assessment (WTA). As discussed above, PFOA, like perfluorooctane sulfonate (PFOS), is widely distributed in the environment. Thus, a detailed evaluation of the environmental effects of PFOA is warranted.

This toxicity assessment systematically summarizes the environmental toxicology of PFOA, as it specifically relates to wildlife. In sequential order, a synthesis of the oral, inhalational, and dermal exposure pathways of PFOA toxicity are described for mammals, birds, amphibians, and reptiles, from which toxicity reference values (TRVs) are derived. Although we have relied significantly on tabulated data as summarized in this report, which provides the reader an opportunity to access the primary or secondary sources of data, we have also summarized the key studies that were used in more detailed assessments and derivation of the TRVs described in this report. It should be noted that an exhaustive description of each study included in this synthesis is beyond the scope of this chapter and certainly not a specific objective of this chapter.

MAMMALIAN ORAL TOXICITY

MAMMALIAN ORAL TOXICITY – ACUTE

Dean and Jessup (1978) reported an oral lethal dose for 50 percent of the animals (LD_{50}) of 680 mg/kg body weight (BW) and 430 mg/kg BW of PFOA for male and female SD (Sprague Dawley) rats, respectively. Other studies have reported an oral LD_{50} of more than 500 mg/kg BW in male Sprague Dawley rats and between 250 and 500 mg/kg BW in female rats (Glaza 1997). In a similar study, an oral LD_{50} of less than 1,000 mg/kg BW was reported for both male and female Sherman-Wistar rats (Gabriel 1976a).

When considering these LD_{50} values in the context of the Hodge and Sterner scale, the data supported the finding that PFOA is moderately toxic after acute oral exposure (Gabriel 1976a). However, since the time that study was completed, several others have provided more detailed dose-response analyses of PFOA in animal models; Table 2.1 provides a summary of the acute oral toxicity studies of PFOA).

TABLE 2.1
Summary of Acute Oral Toxicity for PFOA in Mammals

Test Organism	LD$_{50}$ (mg/kg)	NOAEL (mg/kg-d)	LOAEL (mg/kg-d)	Effects Observed at the LOAEL	Reference
Rats (male)	680	NA	NA	-	Dean and
(female)	430	NA	NA	-	Jessup 1978
Rats (male)	500	NA	NA	-	Glaza 1997
(female)	250–500	NA	NA	-	
Rats (male and female)	1000	NA	NA	-	Gabriel 1976a
Mice (female)	NA	NA	202	Reduced food intake and body weights	Asakawa *et al.* 2008
Mice (sex not stated)	NA	NA	1	Increased absolute and relative liver weights; increased lesion score at all dose levels	Wolf *et al.* 2008
Rats (male)	NA	NA	20	Caused reduced body weight and increased liver and kidney weights	Martin *et al.* 2007
Rats (male)	NA	NA	25	Caused reduced final body weight, overall body weight gain, and decreased food consumption; increased liver weight; increase in interstitial testosterone concentration in testes	Biegel *et al.* 1995
Mice (male)	NA	NA	10 ppm*	Increased liver weights and peroxisome proliferation	Yang *et al.* 2001
Mice (male)	NA	NA	500 ppm	Reduced body weights	Sohlenius *et al.* 1992
Mice (male)	NA	NA	10 ppm	Increased relative liver weights	Xie *et al.* 2002, Yang *et al.* 2000
Mice (male)	NA	NA	200 ppm	Reduced body weight; epididymal and retro-peritoneal fat content were all reduced during treatment; increased liver weights	Xie *et al.* 2003
Mice (male)	NA	NA	200 ppm	Reduced body weights; increased absolute and relative liver weights	Permadi *et al.* 1992

(Continued)

TABLE 2.1 (CONTINUED)
Summary of Acute Oral Toxicity for PFOA in Mammals

Test Organism	LD_{50} (mg/kg)	NOAEL (mg/ kg-d)	LOAEL (mg/ kg-d)	Effects Observed at the LOAEL	Reference
Mice (male)	NA	NA	20 ppm	Increase in liver weights	Qazi *et al.* 2010
Mice (male)	NA	NA	1.6	Increase in liver weights	Qazi *et al.* 2012
Mice (male) (female)	NA	10 ppm 10 ppm (PFOA)	30 ppm 30 ppm (PFOA)	Increased liver weights	Kennedy 1987
Rat (male) (female)	NA	NA 100 ppm	100 ppm 200 ppm	Increased kidney enzyme activities	Kawashima *et al.* 1991
Rat (male) Rat (female)	NA	NA NA	100 ppm 100 ppm	Increased absolute and relative liver weights	Kawashima *et al.* 1994
Rat (male) (female)	NA	NA 200 ppm	12.5 ppm 400 ppm	Increased liver enzyme activities	Kawashima *et al.* 1989
Rat (male)	NA	NA	25 ppm*	Decreased liver glutathione S- transferase activity and increased hepatic concentration of triacylglycerol	Kawashima *et al.* 1995
Mice (male)	NA	NA	1 ppm (PFOA)	Increased liver weights	Eldasher *et al.* 2013
Mice (female)	NA	NA	0.94 ppm (PFOA)	Increased relative liver weights	DeWitt *et al.* 2008
Rats (male)	NA	1 (PFOA)	10 ppm (PFOA)	Increased relative liver weights and serum estradiol concentration	Cook *et al.* 1992

Key: *Dietary concentration provided when no body weight and/or food consumption information available to calculate daily dose.

Legend:

LD_{50}: dose resulting in 50% mortality

NOAEL: no-observed-adverse-effect level

LOAEL: lowest-observed-adverse-effect level

Liver toxicity and reduced body weights were a common finding in many studies following oral administration of PFOA, given orally by gavage or when included in the diet (Wolf *et al.* 2008, Eldasher *et al.* 2013, DeWitt *et al.* 2008, DeWitt *et al.* 2009). Studies reported increased absolute and relative liver weights and increased lesion scores at all dose levels (exposure to 1, 3, or 10 mg/kg-d for 7 days) in mice (Wolf *et al.* 2008). Exposure of adult male mice to PFOA as the free-acid form, at doses of 1 or 3 mg/kg-d for 7 days *via* oral gavage resulted in dose-dependent increases in liver weights without affecting kidney weights (Eldasher *et al*. 2013). In addition, other groups found that dietary PFOA (200 ppm or 30 mg/kg-d; Qazi *et al*. 2012) given to male mice for seven days did not result in any apparent signs of toxicity, such as sores, poor grooming, lethargy, or other behavioral changes (Qazi *et al*. 2012, Xie *et al*. 2003, Yang *et al*. 2000).

In another study (Rigden *et al*. 2015), test groups, each consisting of five male Sprague Dawley rats, were exposed to PFOA at 0, 10, 33, or 100 mg/kg-d for three days and then maintained for an additional four days, with daily body weight measurements and overnight collection of urine specimens. Statistically significant increases in liver weights were noted at all tested doses, and the 10 mg/kg-d dose administered to rats for 3 days was the lowest-observed-adverse-effect level (LOAEL) for the acute toxicological effects of PFOA on the liver (Rigden *et al*. 2015). Furthermore, exposure of male rats to PFOA at 20 mg/kg-d *via* oral gavage for five days decreased mean body weight, but increased both liver and kidney weights after 3 or 5 days of treatment, respectively (Martin *et al*. 2007).

Male rats that were exposed orally to 25 mg/kg-d PFOA for 14 days exhibited reduced final body weight, decreased overall body weight gain, and lower food consumption. Liver weights increased while testicular weights remained unaffected (Biegel *et al*. 1995). In a related study, oral gavage of adult male rats exposed to 1, 10, 25, or 50 mg/kg-d PFOA for 14 days resulted in a dose-dependent decrease in overall body weight gains, whereas relative liver weights increased in rats that had received 10, 25, or 50 mg/kg-d PFOA (Cook *et al*. 1992).

Other research groups found that dietary exposure of male and female rats to PFOA at 15 mg/kg-d for 2 weeks did not affect body weight, but increased absolute and relative liver weights (Kawashima *et al*. 1994). By contrast, dietary exposure of male rats to PFOA at doses of up to 60 mg/kg-d did not significantly affect the rate of body weight gain or food intake. At doses of 7.5 mg/kg-d, exposure to PFOA was associated with increased absolute and relative liver weight (Kawashima *et al*. 1995).

Mammalian Oral Toxicity – Acute: Effects on Hormones

Adult male Sprague Dawley rats were exposed to PFOA at 20 mg/kg-d for one, three, or five days by oral gavage, following which the concentrations of selected hormones, including the hormone precursor cholesterol, testosterone, total thyroid hormones T4 and T3, and free T4 (FT4, thyroxine), were assayed (Martin *et al*. 2007). Following a one-, three-, or five-day PFOA exposure, cholesterol levels had decreased, and total T4, T3, and FT4 levels had decreased (Martin *et al*. 2007). Serum testosterone concentrations had also decreased after a three- or five-day PFOA exposure. PFOA exposure was associated with hepatotoxic effects, including hypocholesterolemia, hepatocellular hypertrophy, hypolipidemia, and peroxisome proliferation (Martin *et al*. 2007).

Mammalian Oral Toxicity – Acute: Endocrine Effects

When adult male rats were dosed with PFOA by oral gavage at 1, 10, 25, or 50 mg/kg-d for 14 days, no significant changes in absolute testicular weights were noted in PFOA-treated rats. Decreases in unit weight of accessory sex organs (consisting of ventral and dorsal lateral prostate, seminal vesicles, and coagulating glands) were seen in rats receiving 25 or 50 mg/kg-d. as compared with controls. No significant differences were found in serum and interstitial testosterone or serum luteinizing hormone levels in any of the PFOA-treated groups compared with the controls. Serum estradiol levels in the groups treated with 10, 25, or 50 mg/kg-d PFOA were higher than those levels found in the *ad libitum* control (Cook *et al.* 1992).

Mammalian Oral Toxicity – Acute: Neurotoxicity

Limited data were available with respect to neurotoxicity or neurodevelopmental effects, following exposure to PFOA. The limited animal model studies do not yet indicate any association between PFOA and possible diverse outcomes. For example, in an NMRI mouse model, the influence of PFOA exposure on habituation and activity patterns was determined on exposure of 10-day-old animals at postnatal day (PND) 10 to a single oral dose of PFOA of 0.58 or 8.7 mg/kg BW, with subsequent evaluation at two and four months of age (Johansson *et al.* 2008, 2009).

Neonatal PFOA exposure at PND 10 caused persistent disturbances in spontaneous behavior of these mice as adults and modified susceptibility of the adult cholinergic system. Changes in spontaneous behavior worsened with age, with the disturbances appearing to be more pronounced in 4-month-old mice as compared with 2-month-old mice. Neonatal PFOA exposure induced irreversible changes in the adult mouse brain that manifested as deranged spontaneous behavior, lack of habituation, and increased susceptibility of the cholinergic system (Johansson *et al.* 2008).

A single PFOA dose at 8.7 mg/kg was given to 10-day-old male mice and the effects were determined for several neurologically active brain proteins, which showed increased hippocampal expression of calcium/calmodulin-dependent protein kinase II (Johansson *et al.* 2009).

In other mouse studies, the offspring of C57BL/6/Bkl dams that were fed PFOA at 0.3 mg/kg-d throughout the entire gestation period had detectable PFOA levels in their brains at birth (Onishchenko *et al.* 2011). When pups were aged five weeks, behavioral studies revealed gender-specific differences in exploratory behavior. In social groups, PFOA-exposed males showed greater activity, whereas PFOA-exposed females were less active than their respective controls. In circadian activity studies, PFOA-exposed males exhibited greater activity than the control males (Onishchenko *et al.* 2011).

Results of an *in vitro* study of hippocampal synaptic transmission and neurite growth in the presence of long-chain perfluorinated compounds showed that 50 or 100 μmol PFOA increased spontaneous synaptic currents and similarly affected neurite growth (Liao *et al.* 2009a, 2009b), which suggested the need for additional studies of the effects of PFAS, including PFOA, on the central nervous system.

MAMMALIAN ORAL TOXICITY – SUB-CHRONIC

Following on from discussions of the acute toxicological effects of PFOA exposure described above, the observed effects of PFOA on the liver is a common indicator of sub-chronic toxicity (see Table 2.2 for a detailed summary of key studies used in the derivation of TRVs).

TABLE 2.2

Summary of Sub-Chronic Oral Toxicity for PFOA in Mammalian Species

			Test Results		
Test Organism	Test Duration/ Period	NOAEL (mg/ kg-d)	LOAEL (mg/ kg-d)	Effects Observed at the LOAEL	Reference
Mice (female)	21 days	1	5	Increased liver weights and liver enzyme activities	Ahmed and Abd Ellah 2012
Mice (male)	21 days	NA	0.49	Increase relative liver weights	Son *et al.* 2008
Mice (male)	21 days	2.64	17.63	Increased lymphocytes in the spleen	Son *et al.* 2009
Mice	21 days	Males: 10 ppm* Females: 10 ppm*	30 ppm 30 ppm	Increased liver weights	Kennedy 1987
Rats (male)	28 days	NA	5	Clinical signs, relative organ weights, lung histopathology	Cui *et al.* 2009
Rats (male)	13 weeks	0.06	0.64	Increase in absolute and relative liver weights and increased activity of liver enzymes	Perkins *et al.* 2004
Mice (female)	GD1–17	NA	0.1	Reduced relative liver weights in offspring	Abbott *et al.* 2007
Mice (female)	GD1–17	NA	5	Reduced number of live offspring, with few surviving to PND7	Abbott *et al.* 2012
Mice (female)	4 weeks	NA	1.0	Reduced uterine weights; increased liver weights; altered mammary gland development	Yang *et al.* 2009
Mice (female)	GD0–18	NA	1.0	Increased relative liver and kidney weights; pathology of the liver and kidney; increased mitosis	Yahia *et al.* 2010
Mice (female)	GD1–17 GD8–17	NA	5	Altered mammary gland development	White *et al.* 2007
Mice (female)	GD7–17 GD10–17 GD13–17 GD15 – 17	NA	3	Altered mammary gland development	White *et al.* 2009
Mice (female)	GD11–16	NA	2	Decreased placental weights, increased resorptions	Suh *et al.* 2011

(Continued)

TABLE 2.2 (CONTINUED)
Summary of Sub-Chronic Oral Toxicity for PFOA in Mammalian Species

Test Organism	Test Duration/ Period	NOAEL (mg/kg-d)	LOAEL (mg/kg-d)	Effects Observed at the LOAEL	Reference
Mice (female)	GD1–PND 21	10	30	Reduced maternal body weights	Hinderliter et al. 2005
Mice (female)	GD1–18	NA	3	Increased liver weights in off-spring, sex-related differences in behavioral responses	Onishchenko et al. 2011
Mice (female)	GD1–17	NA	0.3	Impaired mammary gland development	Macon et al. 2011
Mice (female)	GD1–17	NA	1	Increased liver weight in dams, reduced ossification in offspring, more rapid sexual development	Lau et al. 2006
Mice (female)	GD6–17	NA	0.5	Reduced offspring weight	Hu et al. 2010
Mice (female)	GD1–17	NA	3	Reduced pup survival and histopathological changes	Albrecht et al. 2013
Mice (female)	4 weeks	NA	5	Increased progesterone levels in females in pro-estrus or estrus	Zhao et al. 2010
Mice (female)	4 weeks	NA	2.5	Impaired mammary gland development and delayed vaginal opening	Zhao et al. 2012
Rats (male and female)	26 weeks	NA	1,000 ppm	Induction of liver enzymes	Uy-Yu et al. 1990a

Key: *Dietary concentration provided when no body weight and/or food consumption information available to calculate daily dose.

Legend:
NOAEL: no-observed-adverse-effect level
LOAEL: lowest-observed-adverse-effect level
NA: not applicable

For example, mice treated orally with PFOA for 21 days exhibited increased liver weights at 5 or 10 mg/kg-d, although no effects were observed at a dose of 1 mg/kg. There were significant increases in hepatic total glutathione concentration at doses of 5 or 10 mg/kg body weight per day, and hepatic catalase activity was increased at a PFOA dose of 5 mg/kg-d body weight. Hepatic glutathione reductase activity was also increased at a PFOA dose of 10 mg/kg-d (Ahmed and Abd Ellah 2012). Other groups have investigated the effects of exposing male mice to PFOA in drinking water for 21 days at concentrations equivalent to 0.49, 2.64, 17.63, or 47.21 mg/kg-d (Son et al. 2008, 2009). It was found that exposure of mice to PFOA at a dose of 17.63 mg/kg-d body weight caused decreased food and water consumption. In addition, body weights were reduced in animals that were exposed to PFOA at doses equivalent to 17.63 or 47.21 mg/kg body weight per day (Son et al. 2008, 2009).

Furthermore, it was found that relative liver weights were increased at doses of PFOA as low as 0.49 mg/kg body weight per day; however, relative kidney weights were unchanged. Plasma levels of blood urea nitrogen and creatinine were unaffected, observations that suggested no kidney toxicity. Changes in liver histopathology were examined, with marked hepatocytomegaly and an acidophilic cytoplasm being observed in PFOA-treated mice (Son et al. 2008). At high doses of PFOA (at $\geq$ 17.63 mg/kg-d), diffuse hepatic damage, following multi-focal coagulation and liquefaction necrosis, were also seen. By contrast, no changes in kidney histopathology were noted (Son et al. 2008).

Yang et al. (2009) also studied exposure of female BALB/c and C57BL/6 mice to PFOA. Mice were exposed once daily, for five days per week, for four weeks by oral gavage at doses of 1, 5, or 10 mg/kg body weight. Exposure to lower doses of PFOA did not affect animal body weights. However, groups of both strains of mice that received the highest dose of 10 mg/kg-d showed decreased body weights towards the end of the study. In addition, dose-dependent increases in absolute and relative liver weights were seen in both strains of mice receiving the lowest PFOA dose of 1 mg/kg-d (Yang et al. 2009).

In another study, pregnant ICR mice were dosed with PFOA at 1, 5, or 10 mg/kg via oral gavage from gestational day (GD) 0 to 17 and 18 for prenatal and postnatal evaluations, respectively (Yahia et al. 2010). Five to nine dams per group were sacrificed on GD18 for prenatal evaluations. Although no maternal deaths from PFOA exposure were noted, liver weight was shown to increase in a dosage-dependent manner, including evidence of hepatocellular hypertrophy, necrosis, increase mitosis, and mild calcification at 10 mg/kg-d. In addition, total body weight was reduced in mice exposed to at least 5 mg/kg PFOA per day, with both relative liver and kidney weights increasing at 1 mg/kg-d dosage and greater. The kidney also showed slight hypertrophy in the outer medulla and proximal tubular cells in all treatment groups (Yahia et al. 2010).

Exposure of male rats to PFOA in the diet for 13 weeks at nominal concentrations that were equivalent to 0.06, 0.64, 1.94, or 6.5 mg/kg-d caused one death at the 6.5 mg/kg dose (Perkins et al. 2004). However, there were no clinical signs of toxicity at any dietary PFOA concentration. Body weight gains were reduced in animals that were dosed with 6.5 mg/kg-d PFOA. In addition, PFOA caused no major differences in the incidence of gross pathological findings between controls and treated animals; however, at doses of 0.64 mg/kg-d or greater, PFOA exposure was associated with increased absolute and relative liver weights after 4, 7, and 13 weeks of feeding. There was minimal to mild hepatocellular hypertrophy in the livers of rats that were dosed at 0.64, 1.94, or 6.5 mg/kg-d after 4, 7, and 13 weeks of feeding, each of which was reversible following a recovery period. In addition, dietary exposure of rats to PFOA was associated with a significant increase in hepatic palmitoyl CoA oxidase activity at doses of 0.64 mg/kg-d or higher, which peaked by 7 weeks of exposure. The no-observed-adverse effect level (NOAEL) was determined to be 0.06 mg/kg-d, which was based on increases in absolute and relative liver weight and hepatocellular hypertrophy, and the LOAEL was determined to be 0.64 mg/kg-d (Perkins *et al.* 2004).

Cui *et al.* (2009) explored the toxicological effects of PFOA in male Sprague Dawley rats. Oral doses of PFOA at 5 mg/kg-d for 28 days was associated with reduced activity, decreased food uptake, cachexia, and lethargy, starting in the third week of exposure. Female rats were treated during pregnancy *via* gavage with 100 mg/kg-d PFOA for ten days. Female rats dosed with PFOA exhibited wet abdomens, which began in the pelvic area, had blood-stained tears and nasal discharge, and were unkempt. In addition, a PFOA dose of 20 mg/kg-d was associated with increased sensitivity to external stimuli. An oral dose of 20 mg/kg-d for 28 days was associated with reduced weight gain – although food consumption remained unaffected. Relative liver, kidney, and testis weights were higher in rats exposed to PFOA at 20 or 100 mg/kg-d as compared untreated control animals. No gross liver pathology was observed in rats that had received 5 mg/kg-d; however, tumefaction and dark coloration patterns were observed in the livers from rats that were treated with 20 mg/kg-d (Cui *et al.* 2009). Liver histopathology revealed cytoplasmic vacuolation, focal or flake-like necrosis, and hepatocellular hypertrophy at both PFOA doses of 5 and 20 mg/kg-d. PFOA exposure was also associated with fatty degeneration, angiectasis, and congestion in the hepatic sinusoid or central vein, and the presence of acidophil lesions. Lung histopathology revealed pulmonary congestion as well as focal or diffusely thickened epithelial walls in rats that received a PFOA dose of 5 mg/kg-d. Upon examination of the kidneys, this analysis showed no obvious pathology at 5 mg/kg-d. However, turbidity and tumefaction of the proximal convoluted tubule epithelium was observed in rats that had received 20 mg/kg-d, which was accompanied by mild clinically associated symptoms, including congestion in the renal cortex and medulla and enhanced cytoplasmic acidophilia (Cui *et al.* 2009).

Male and female mice given diets that contained 0.01, 0.03, 0.1, 0.3, 1, 3, 10, or 30 ppm PFOA for 21 days exhibited increased liver weights at the 30 ppm dose. Body weights of mice fed up to and including 30 ppm PFOA were similar to those of the controls, and there was no evidence of any unusual clinical pathology in these mice (Kennedy 1987). Additionally, other researchers found that exposure of animals to PFOA rarely caused mortality. However, exposure of female rats to only a single dose during pregnancy *via* gavage with 100 mg/kg-d PFOA for ten days resulted in three deaths from the 25 animals that were treated with PFOA (Staples *et al.* 1984). During the dosing period, the PFOA-treated group of pregnant dams that were gavage with a single dose of 100 mg/kg-d consumed less feed than did the non-PFOA exposed control group (Staples *et al.* 1984).

In rhesus macaque (*Macaca mulatta*) models (n = 2 per gender per treatment group), PFOA was administered at 3, 10, 30, or 100 mg/kg-d by oral gavage for 90 days (Goldenthal 1978). Animals were observed twice daily, with body weights recorded every week. In addition, blood and urine samples were collected once during a control period, and at one and three months for measurements of hematology, clinical chemistry, and urinalysis. Organs and tissues from animals sacrificed at the end of the study, and from animals that died during treatment were weighed and examined for gross pathology and histopathology. At a dose of 100 mg/kg-d, all of the animals died between weeks 2 and 5 of the study. Additionally, three animals from the 30 mg/kg-d group died during the study, and, beginning in week 4, all four animals showed slight to moderate, and sometimes severe, decreased activity. None of the monkeys in the 3 or 10 mg/kg-d groups died during the conduct of this study (Goldenthal 1978).

Changes in body weight between animals from the 3 and 10 mg/kg-d groups and the control group were similar. In addition, monkeys from the 30 and 100 mg/kg-d groups lost body weight, beginning after week 1. At the end of the study, this weight loss was statistically significant for one male animal that survived treatment in the 30 mg/kg-d group and was reflected in body weight (2.30 kg *versus* 3.78 kg for the control). The results of the urinalysis, and hematological and clinical chemistry analyses were comparable for the control and the 3 and 10 mg/kg-d groups at one and three months. At necropsy, there were significant decreases in the absolute heart and brain weights and relative liver weights in female animals exposed to 10 mg/kg-d. At 3 mg/kg-d, the relative pituitary weight in males was markedly increased, compared with those from the control group; however, these changes were not accompanied by changes in morphology (Goldenthal 1978).

In animals that died, one male and two females from the 30 mg/kg-d group and all four animals from the 100 mg/kg-d group displayed significant diffuse lipid depletion in the adrenal glands. All males and females from the 30 and 100 mg/kg-d PFOA dose groups displayed slight to moderate bone marrow hypocellularity and moderate atrophy of the splenic lymphoid follicles. The one male that survived in the 30 mg/kg-d group until terminal sacrifice had slight to moderate bone marrow hypocellularity and moderate atrophy of splenic lymphoid follicles (Goldenthal 1978b). From observations reported in this study, the male LOAEL was 3 mg/kg-d, which was based on increased relative pituitary weight, although a NOAEL could not be determined. From observations that were based on decreased heart and brain weights, the female LOAEL was 10 mg/kg-d and the NOAEL was 3 mg/kg-d (Goldenthal 1978). Results from the above studies are summarized in Table 2.2.

Mammalian Oral Toxicity –Sub-Chronic: Developmental Toxicity

When pregnant mice were exposed to PFOA from GD 1–17, maternal weight was unaffected in mice that received 3 mg/kg-d (Albrecht *et al.* 2013, Onishchenko *et al.* 2011) or 5 mg/kg-d (Abbott *et al.* 2007, 2012, Fenton *et al.* 2009). Similarly, exposure of pregnant female mice to 5 mg/kg-d PFOA for GD 1–17, GD 8–17, or GD 12–17 did not affect gains in maternal body weight (White *et al.* 2007). However, PFOA treatment at a dose of 25 mg/kg-d for GD 11–16 reduced gains in maternal body weight (Suh *et al.* 2011) and PFOA treatment for GD 1–17 caused decreased weight gains at doses of 25 or 40 mg/kg-d (Lau *et al.* 2006). Furthermore, PFOA treatment throughout pregnancy at a dose of 30 mg/kg-d caused a 10 percent decrease in maternal body weight (Hinderliter *et al.* 2005). Exposure of female mice to PFOA doses up to 1 mg/kg-d from PND 18–20 did not affect body weight (Dixon *et al.* 2012). Some of the PFOA-mediated developmental effects appeared to be associated with expression of the peroxisome proliferator-activated receptor-alpha (PPARα) transcription factor (Abbott *et al.* 2007).

When pregnant mice were dosed with PFOA from GD 1–17, maternal weight, embryonic implantation, or the number or weight of pups at birth were all unaffected at a dose of 5 mg/kg-d. Additionally, at the 5 mg/kg-d PFOA dose, the incidence of full litter resorption was increased – an effect that was concluded to be unrelated to PPARα expression. Reduced neonatal survival at a PFOA dose of 0.6 mg/kg-d and delayed eye opening at 1 mg/kg-d was associated with PPARα expression (Abbott *et al.* 2007). Similarly, postnatal survival of pups that were dosed with PFOA at 0.6 mg/kg-d and body weight changes at maternal doses of 1 mg/kg-d was also associated with PPARα expression. Pups exhibited increased liver weights in the lowest dose group of 0.1 mg/kg-d (Abbott *et al.* 2007) and when the pregnant dams were treated throughout pregnancy with only one dose of PFOA at 0.3 mg/kg-d (Onishchenko *et al.* 2011).

Metabolic disruption, mediated by transcriptional regulation of fatty acid bio-synthesis, metabolism, and β-oxidation, and glucose metabolism *via* peroxisome proliferator-activated receptors, the constitutive androstane receptor and pregnane X-receptor activation, is thought to be a contributing factor to the developmental toxicity of PFOA (Lau 2012).

Exposure to oral doses of PFOA as low as 1 mg/kg for four weeks disrupted development of female reproductive systems during gestation. PFOA treatment caused mammary gland growth inhibition at the 5 and 10 mg/kg doses by reduced ductal length, as well as decreased numbers of terminal end buds and stimulated terminal ducts (Yang *et al.* 2009). Dams treated during GD 8–17 and GD 1–17 exhibited significant visible delays in epithelial differentiation, and developmental scores for mammary glands, compared with controls, and these glands morphologically resembled the glands of dams studied days earlier in lactation (White *et al.* 2007).

When pregnant female mice were exposed to 5 mg/kg-d, periods of GD 7–17, GD 10–17, GD 13–17, GD 15–17, or even the shortest treatment duration (GD 15–17) were sufficient to consistently lower mammary gland development scores in offspring at PND 29 and 32, as compared with controls (White *et al.* 2009). Mammary glands from PFOA-treated mice exhibited histological features of delayed epithelial growth, which were similar to those seen in rats that were neonatally exposed to endocrine-disrupting compounds. By visual observation, mammary glands of PFOA-exposed mice displayed aberrant morphology and were thus given lower developmental scores than controls. Reduced developmental score were also given for mammary glands that were examined for all PFOA-dosed groups (Macon *et al.* 2011). Exposure to PFOA, regardless of whether it was administered during gestation or *via* lactation, was associated with reduced mammary gland developmental scores (White *et al.* 2009).

Exposure to oral doses of PFOA as low as 1 mg/kg-d for four weeks disrupted development of female reproductive systems during gestation. PFOA treatment caused inhibition of mammary gland growth at the 5 and 10 mg/kg-d doses by reduced ductal length, as well as by decreased numbers of terminal end buds and stimulated terminal ducts (Yang *et al.* 2009). Dams treated at GD 8–17 or GD 1–17 exhibited significant visible delays in epithelial differentiation and developmental scores for mammary glands, compared with controls, and these glands morphologically resembled glands of dams which were days earlier in lactation (White *et al.* 2007).

In addition, when pregnant mice were exposed to 5 mg/kg-d PFOA at GD 7–17, GD 10–17, GD 13–17, or GD 15–17, then even the shortest treatment duration (i.e., GD 15–17) consistently lowered mammary gland development scores in offspring at PND 29 and 32, as compared with controls (White *et al.* 2009). Mammary glands from PFOA-treated groups exhibited histological characteristics of delayed epithelial growth that were similar to rats that had been neonatally exposed to endocrine-disrupting compounds. On visual observation, mammary glands of PFOA-exposed mice displayed aberrant morphology, and consequently were given lower developmental scores as compared to controls. Furthermore, reduced developmental scores were given for mammary glands in all groups (Macon *et al.* 2011). Exposure to PFOA, regardless of whether it was given during gestation or lactation, was associated with reduced mammary gland developmental scores (White *et al.* 2009).

Yang *et al.* (2009) administered PFOA to 21-day-old female BALB/c mice at 0, 1, 5, or 10 mg/kg-d by oral gavage for five days per week for four weeks. The goal of this study was to determine the effects of peripubertal exposure to PFOA on puberty and mammary gland development. A significant decrease in body weight was found following exposure of mice to 10 mg/kg-d PFOA. In addition, the mammary glands of female mice exposed to 5 or 10 mg/kg-d exhibited reduced ductal length, decreased numbers of terminal end buds, and decreased stimulated terminal ducts, as compared with the mammary glands of their normal control counterparts.

Absolute and relative liver weights were increased at all PFOA doses tested in this model (Yang *et al.* 2009). Additionally, the absolute and relative uterine weights were reduced in all treated mice, compared with the uterine weights of controls. Mean age at vaginal opening was greater in those mice treated with 1 mg/kg-d and did not occur in mice that were treated with PFOA at doses of 5 or 10 mg/kg-d. Under these observations, the LOAEL was determined to be 1 mg/kg-d, which was based on delayed vaginal opening, increased liver weight, and decreased uterine weight. For these measured outcomes, a NOAEL could not be determined (Yang *et al.* 2009).

This same research group also dosed 21-day-old female C57BL/6 mice as described above for the studies conducted in BALB/c mice (Yang *et al.* 2009). Similar trends in body weight effects in both strains were found on exposing animals to PFOA at 10 mg/kg-d, which induced a reduction in body weights. Although ductal length remained unaltered, 5 mg/kg-d PFOA stimulated the mammary glands with a significant increase in terminal end-bud numbers and stimulated terminal ducts. On the other hand, mammary gland development was inhibited in mice that were dosed at 10 mg/kg-d, an effect that was characterized by a lack of terminal end buds, an absence of stimulated terminal ducts, and suppression of ductal growth (Yang *et al.* 2009).

Furthermore, all treated mice showed increased absolute and relative liver weights. The absolute and relative uterine weights were increased in C57BL/6 mice dosed with 1 mg/kg-d PFOA and decreased in mice that were dosed with 10 mg/kg-d. There was no difference from the control values in terms of uterine weights of mice treated with 5 mg/kg-d. Vaginal opening was delayed in C57BL/6 mice dosed with 5 mg/kg-d and did not occur at all in mice dosed with 10 mg/kg-d. The LOAEL was 1 mg/kg-d based on increased liver and uterine weights, and no NOAEL was established (Yang et al. 2009).

Young female mice aged PND 18–20, were dosed with between 0.1 and 1 mg/kg-d PFOA, receiving it as the PFOA salt,. This dosing strategy provoked slight increases in uterine wet weight only at a dose of 0.1 mg/kg-d (Dixon et al. 2012). Mice that were treated with PFOA had minimal histopathological changes in the uterus, cervix, and vagina. In the uterus, there was minimal to mild endometrial and myometrial edema, in addition to minimal thickening (i.e., hyperplasia) of the uterine mucosal and endometrial glandular epithelia and smooth muscle layers. There was also focal minimal stromal edema of the cervix, and the vagina had focal areas of mucification. However, cornification and squamous hyperplasia of the cervical and vaginal mucosae were largely absent (Dixon et al. 2012).

Dosing pregnant mice with PFOA at 1, 5, or 10 mg/kg-d via oral gavage at GD 0–18 was associated with reduced fetal body weights at doses of 5 or 10 mg/kg-d (Yahia et al. 2010). However, prenatal survival was unaffected and administration of PFOA at any of the doses did not cause any gross abnormalities. Developmental effects, as illustrated by an increased incidence of observed cleft sternum, delayed ossification of phalanges, and delayed eruption of incisors, were seen at a PFOA dose of 10 mg/kg-d. There was also evidence of reduced survival of neonates at 5 and 10 mg/kg (Yahia et al. 2010). Similarly, Suh et al. (2011) found decreased fetal weights in time-mated mice that were administered at doses of 2, 10, or 25 mg/kg-d PFOA by oral gavage at GD11–16. Decreased placental weights and an increased number of resorptions were also found in all PFOA-treated groups. Additionally, decreased numbers of live fetuses were seen in groups that were dosed with PFOA at 10 or 25 mg/kg-d. However, there were no differences seen in the number of implanted embryos.

In a separate study, exposure of pregnant mice to 5 mg/kg-d of PFOA at GD 1–17, GD 8–17, or GD 12–17 did not affect the mean number of implantation sites, live pups born, or embryonic/fetal loss rates. On PND 1, reduced body weights were found among prenatally PFOA-exposed pups in a time-dependent manner, with the effect on pup body weight being sustained throughout the period of lactation in all PFOA-treated groups (White et al. 2007).

Other research showed increased liver weights at all doses tested in response to oral gavage of pregnant female mice at GD 1–17, when PFOA was administered at dosages in the range 1–40 mg/kg-d (Lau et al. 2006). The number of dams with full litter resorptions increased at 5 mg/kg-d dose and above, with all of the dams displaying full litter resorptions at 40 mg/kg-d. Furthermore, the number of live fetuses in live litters was also reduced at 18 days, with increasing prenatal losses seen at 20 mg/kg-d. Reduced fetal body weights were also seen at 18 days at 20 mg/kg-d and reduced ossification was seen at various sites at a PFOA dose equal to or greater than 1 mg/kg-d. There was also a slight delay in parturition at PFOA doses of 3, 10 or 20 mg/kg-d, but not at 5 mg/kg-d, for reasons that remained unclear (Lau *et al.* 2006).

Exposure of pregnant female mice to 3 mg/kg-d PFOA at GD1–17 did not affect average gravid uterine weight, the number of implantations per dam, the number of resorptions per litter, or the average number of fetuses per litter (Albrecht *et al.* 2013). In addition, exposure of dams to 3 mg/kg-d (body weight) PFOA did not affect average crown-to-rump length, average body weight, or the number of live or dead fetuses per litter. Additionally, there was no delay in parturition in response to exposure to PFOA (Albrecht *et al.* 2013). Exposing female rats to 100 mg/kg-d PFOA by oral gavage at GD 6–15 did not affect pregnancy maintenance or the number of resorptions. Fetal body weight was also unaffected, and malformations were not seen in this study (Staples *et al.* 1984).

Other groups provided pregnant female mice with PFOA at GD 6–17 in their drinking water at doses that were equivalent to 0.5 or 1 mg/kg-d, and found that the number of dams delivering viable pups was unaffected (Hu *et al.* 2010). Additionally, when studying the delivered pups, it was found that the litter size, sex ratio, and organ weights of the adrenal gland, liver, spleen, and thyroid were unaffected (Hu *et al.* 2010).

Impacts from PFOA exposure during gestation were subsequently seen in the delivered offspring. When female mice were exposed to PFOA at a dose of 3 or 5 mg/kg-d during gestation and/or lactation, and the offspring from these dams were cross fostered, all PFOA-exposed groups exhibited lower body weights than the controls at PND 22. However, this was not seen in mice that were exposed to PFOA at 3 mg/kg-d during lactation. All PFOA-exposed groups had increased relative liver weights at weaning (White *et al.* 2009). Pups born to dams, that were administered PFOA at GD 6–17 at doses of 0.5 or 1.0 mg/kg-d *via* the drinking water, weighed less than control group animals at PND 2 only. In addition, pups that were born to dams dosed with 1.0 mg/kg-d PFOA weighed less than their control group counterparts through to PND 14 (Hu *et al.* 2010).

In a separate study, Lau *et al.* (2006) reported reduced neonatal body weights at or around PND 10, following the administration of PFOA to pregnant dams at 5, 10, or 20 mg/kg-d. This group also found reduced neonatal survival and delayed eye opening at a dose of 5 mg/kg-d or above. Sexual development was accelerated at a PFOA dose of 1 mg/kg-d or above (Lau *et al.* 2006). The average weight per pup per litter and the average gain in pup weight from PND 0 through to PND 20 were unaffected on exposure of the dams to 3 mg/kg-d at GD 1–17. The date of onset of eye opening did not change in this group. Pup survival from PND 0 through PND 20 decreased, compared with the zero-PFOA controls, in the 3 mg/kg-d PFOA exposure group. In both dams and pups, there was evidence of altered histopathology – an observation that included peroxisome proliferation (Albrecht *et al.* 2013).

In another full gestation study, pregnant CD-1 mice were orally administered PFOA at GD 1–17 at doses of up to 3 mg/kg-d (Macon *et al.* 2011). Resultant male and female pups did not show any changes in body weight; however, absolute and relative liver weights were increased in both male and female pups across all groups, the significant effects lasting to PND 28 in females and PND 42 in males.

Furthermore, decreased absolute brain weights were seen at PND 63 in male pups from dams that had received 1.0 or 3.0 mg/kg-d. In addition, from PND 4 to PND 7, increased absolute liver weights were seen in pups born to PFOA-exposed dams at a dose of 1.0 mg/kg-d, compared with controls. Similarly, from PND 4 to PND 14, increased relative liver weights were seen in offspring that were born to PFOA-dosed (1 mg/kg-d) dams (Macon *et al.* 2011).

Exposing pregnant mice to PFOA at doses up to 5 mg/kg-d across the full gestational period at GD1–GD17 *via* oral gavage shortened pup lifespans (Hines *et al.* 2009). At the age of 51 weeks, no mortality was seen in the controls; however, mortality rates of 20%, 10%, 36%, and 6% were seen in the offspring of dams following dosing at 0.01, 0.1, 0.3, or 1 mg PFOA/kg-d, respectively. By 76 weeks, there was a 40% mortality rate in controls, and mortality rates of 32%, 63%, 60%, and 44% in the 0.01, 0.1, 0.3, and 1 mg PFOA/kg-d groups, respectively. No significant differences in mortality rates were seen between the control group and any of the treatment groups at specific times in later life or in survival rates across time (Hines *et al.* 2009). Similarly, other research groups showed that the number of live fetuses and the number of implantation sites at GD18, or the number of live-born pups at PND1 all remained unaltered after a one-time oral gavage dosing of pregnant mice with up to 5 mg/kg PFOA at GD17. In addition, pup body weight, and absolute and relative liver weights remained unaltered (Fenton *et al.* 2009).

Abbott *et al.* (2012) determined that exposure of pregnant mice to 5 mg/kg-d from GD1–17 did not affect the number of embryos that were implanted up to GD14. However, the number of live pups at birth and survival to PND7 was lower in PFOA-exposed litters *in utero*. Abbot *et al.* (2012) proposed that the metabolic disruption that was caused by PFOA exposure was likely mediated through disruption of the expression of genes that play critical roles in fatty acid biosynthesis, metabolism, beta-oxidation, and glucose metabolism. PFOA-mediated metabolic disruption might contribute to the poor postnatal survival and growth that was observed in CD-1 neonates following *in utero* exposure to PFOA.

Effects on neurological development were investigated by feeding pregnant C57BL/6 bkl mice the equivalent of 0.3 mg/kg-d PFOA in their food throughout pregnancy (Onishchenko *et al.* 2011). From this study, it was determined that, at the time of birth, there were no differences in offspring body or brain weights between groups. In addition, PFOA exposure did not significantly affect locomotor activity in either male or female mice. Gender-specific differential activities in circadian rhythm were seen during the first hour of the test when measured by the TraffiCage system. It was found that PFOA-exposed males were more active, whereas PFOA-exposed females showed decreased activity, compared with the controls. In addition, PFOA-treated males were more active during the light and dark phases, but treated females displayed similar activities to those seen in the control group. No differences existed in anxiety-related behavior as measured in the elevated plus-maze, or in an analysis of a forced swimming test. PFOA exposure did not significantly affect muscle strength in either male or female mice, and did not affect motor coordination in male mice, whereas PFOA-exposed females performed more poorly than did the controls (Onishchenko *et al.* 2011).

In a rat model, pregnant animals that were dosed with PFOA at up to 30 mg/kg-d from GD 1 to PND 21, showed no evidence of maternal, fetal, or pup mortality, or PFOA-related clinical observations at any of the dose levels tested (Hinderliter et al. 2005). Subsequently, the numbers of fetuses in the gestational portion of the study appeared normal in all dosed groups. Pup survival and weight during lactation were also comparable across groups, and PFOA-related pup mortality was not seen during lactation (Hinderliter et al. 2005).

Mammalian Oral Toxicity – Subchronic: Endocrine Effects

Exposure of 21-day-old female mice to 5 mg/kg PFOA via oral gavage 5 days a week for 4 weeks did not affect total serum estradiol levels, compared with control mice or those treated with PFOA in proestrus or in stages of the estrus cycle. PFOA treatment increased serum progesterone levels by approximately threefold in mice that were in proestrus or estrus stages. These results demonstrate that the stimulatory effect of peripubertal PFOA treatment on mammary gland development in C57BL/6 mice was independent of the expression of PPARα (Zhao et al. 2010).

In a related study, female mice receiving oral gavage doses of 2.5 or 7.5 mg/kg PFOA for 5 days a week for 4 weeks did not show significantly altered body weight when exposed to PFOA at 2.5 mg/kg-d (Zhao et al. 2012). However, mice treated with 7.5 mg/kg-d showed decreased body weight, but only demonstrated this effect during the final week (i.e., the 4th week) of PFOA treatment. Exposure of mice to 2.5 or 7.5 mg/kg-d of PFOA inhibited mammary gland growth, as evidenced by reduced ductal length, decreased numbers of terminal end buds (TEBs), and stimulated ducts (STDs). In addition, the mean age at which vaginal opening occurred was significantly later or even absent in PFOA-treated mice. These results indicated that PFOA exposure delayed and even disrupted ovarian function in PFOA-treated mice (Zhao et al. 2012).

In the Hines et al. (2009) study, low doses of PFOA given during gestation to CD-1 mice not only increased weight gain, but also increased serum levels of insulin and leptin of the offspring in mid-life. In a complementary study (Quist et al. 2015); the mature animals from the Hines et al. (2009) study were investigated. It was determined that there was no dose-dependent impact of PFOA on serum leptin concentration in the CD-1 offspring which had been exposed during gestation across a dose range of 0–1 mg/kg-d when assessed at PND 91. However, in the mouse group given a high-fat diet (HFD) and not fasted before serum collection, a PFOA dose-dependent decrease in leptin level was found (Quist et al. 2015). This study also found that the fat content of the diet and the time of serum collection were critical variables, that need further exploration, because other mice on the HFD which were fasted for 4 hours prior to serum collection in the same study did not show a dose-dependent effect on leptin levels (Quist et al. 2015). Furthermore, diet might adversely influence the risk associated with PFOA exposures, since several animal studies have demonstrated an increased risk of liver steatosis in insulin resistance animal models that were fed an HFD (Quist et al. 2015, Tan et al. 2013, Hines et al. 2009).

Mammalian Oral Toxicity – Sub-Chronic: Immunotoxicity

In a separate study, non-significant decreases in spleen and thymus weights were found in mice that had previously been treated with PFOA. Exposure to PFOA in the drinking water for 21 days also altered specific lymphocyte sub-populations in the spleen and thymus of exposed mice. For example, lymphocyte numbers were increased in the spleen at calculated PFOA doses of 17.63 and 47.21 mg/kg-d, with increased lymphocyte numbers in the thymus at a dose of 47.21 mg/kg-d. Histopathology of the spleen and thymus showed altered morphology at a dose of 47.21 mg/kg-d. The expression of proinflammatory cytokines was altered in the spleen and thymus at a dose of 47.21 mg/kg-d. Based on the increased frequency of lymphocytes in the spleen, the LOAEL for this study was 17.63 mg/kg-d, and the NOAEL was 2.64 mg/kg-d (Son *et al.* 2009).

In complementary studies (DeWitt *et al.* 2008), the immunotoxic effects of PFOA were explored by determining humoral and cell-mediated immune responses. This study exposed groups (n = 8 per group; five groups) of C57BL/6J mice to a single oral gavage dose of PFOA in distilled water at 30 mg/kg-d for ten continuous days. Immediately following this exposure, half of the total number of mice continued to be exposed to PFOA from day 11–15, whereas the remaining animals were dosed with only distilled water from day 11–15 (DeWitt *et al.* 2008). Immune responses were determined following the immunization of 16 animals per group with sheep red blood cells (SRBC) on day 11, and eight mice per group were administered bovine serum albumin (BSA). Animals were euthanized on day 11 (i.e., 1 day post-exposure) or on day 31 (i.e., 15 days post-exposure) to determine body and organ weights, serum-borne IgM levels, delayed-type hypersensitivity (DTH) reactions to BSA, and serum-borne IgG levels following a boosting immunization with SRBC (DeWitt *et al.* 2008).

Findings associated with measurements of body and organ weights in the DeWitt *et al.* (2008) study agreed with those previously reported (Yang *et al.* 2001). Although body weight was decreased, compared with the control mice, from days 8–11 for both PFOA-treated groups, and on day 16 for mice in the constant exposure group, it was found that, by day 31, body weight of the animals was comparable between groups. In addition, relative liver weight was increased in both PFOA-treated groups on days 16 and 31. Furthermore, decreased absolute and relative weights of both the spleen and thymus were found in both groups of PFOA-exposed animals, as compared with control groups, on day 16. By day 31, similar thymus and spleen weights were found when comparing control and treated mice (DeWitt *et al.* 2008). Additionally, by post-exposure day 1, IgM levels in SRBC-immunized mice had decreased by 20 percent, compared with the control mice; however, the levels of SRBC-specific IgG and delayed-type hypersensitivity (DTH) reactivity were similar between PFOA-exposed and non-exposed, control animals (DeWitt *et al.* 2008).

In addition, two complementary dose-response studies were conducted as part of the above project (DeWitt *et al.* 2008), which enabled the derivation of both a NOAEL and a LOAEL for immunotoxic end-points. In the first dose-response experiment, female C57Bl/6N mice (n = 16 per group) were exposed for 15 days to PFOA in their drinking water at doses of 0, 3.75, 7.5, 15, or 30 mg/kg-d. Body weight was decreased from days 8–16 at 30 mg/kg-d, and on day 16 at 15 mg/kg-d. All doses of PFOA increased liver weights at days 16 and 31. At PFOA doses of 15 mg/kg-d or higher, absolute and relative weights of the spleen and thymus were decreased (DeWitt et al. 2008).

The SRBC-specific IgM response was reduced in a PFOA dose-dependent manner at 3.75 mg/kg-d or higher. By contrast, the SRBC-specific IgG response was somewhat increased but only at the 3.75 and 7.5 mg/kg-d doses of PFOA, increases that were found not to be statistically significant on comparison to their controls. Furthermore, none of the PFOA doses affected DTH in this experiment. Based on increased liver weights, as well as the decreased IgM and increased IgG responsiveness to SRBC immunization described here, the LOAEL for PFOA was 3.75 mg/kg-d (DeWitt et al. 2008).

Additionally, the above end points were largely confirmed by a second dose-response experiment, wherein an immunological LOAEL of 3.75 mg/kg-d was derived. Furthermore, the immunological NOAEL was derived to be 1.88 mg/kg-d. Benchmark dose (BMD) analysis of IgM responses gave a lower bound 95% confidence limit of 1.75 mg/kg-d on a BMD ($\pm$ 1SD, standard deviation about the mean) of 3.06 mg/kg-d. In addition, all doses of PFOA tested resulted in significant increases in the liver weight of treated animals on days 16 and 31. The LOAEL, based on observations of increased liver weight, was 0.94 mg/kg-d PFOA (DeWitt et al. 2008).

Mammalian Oral Toxicity – Sub-Chronic: Effects on Enzymes

Exposure to PFOA in the diet of male rats for 13 weeks at nominal concentrations equivalent to 0.06, 0.64, 1.94, or 6.5 mg/kg-d caused statistically significant increases in hepatic palmitoyl CoA oxidase activity (which is a functional indicator of peroxisome activity) at dose levels of 0.64, 1.94, and 6.5 mg/kg-d (Perkins et al. 2004). Furthermore, at the lower dose of 1.94 mg/kg-d, palmitoyl CoA oxidase activity was found to be 2.8-fold higher than that seen for pair-fed controls (Perkins et al. 2004).

Exposure of male rats to 1,000 ppm PFOA in the diet induced marked increases in the activity of liver peroxisomal beta-oxidation pathways and carnitine acetyltransferase at both two and 26 weeks of treatment (Uy-Yu et al. 1990a). Female rats showed only slight responses, with the activity of palmitoyl CoA oxidase being significantly induced only after 26 weeks of PFOA treatment, when compared with age-matched female control rats. In addition, carnitine acetyltransferase activity was increased (compared with the control rats) in female rats at both two and 26 weeks of PFOA treatment, yet the activities of both palmitoyl CoA oxidase and carnitine acetyltransferase were significantly lower in PFOA-treated female rats than in PFOA-treated male rats (Uy-Yu et al. 1990a).

In similar studies, male and female rats were exposed to PFOA at a dose of 1,000 ppm in the diet, for 2, 22, or 26 weeks. Following exposure, several liver enzyme activities were assayed, including stearoyl-CoA desaturase, 1-acylglycerophosphocholine acetyltransferase, and peroxisomal β-oxidation (Uy-Yu et al. 1990b). Exposure of male rats to PFOA increased the activities of stearoyl-CoA desaturase, 1-acylglycerophosphocholine acetyltransferase, and peroxisomal β-oxidation, following two weeks of treatment. Furthermore, increased activities of 1-acylglycerophosphocholine acetyltransferase and peroxisomal β-oxidation were sustained throughout the 22- and 26-week periods of the long-term treatment (Uy-Yu *et al.* 1990b).

Induced activities of microsomal 1-acylglycerophosphocholine acetyltransferase and peroxisomal β-oxidation activity in male rats were highly positively correlated. Hepatic responses to PFOA remained consistent throughout the 26-week period of exposure to PFOA. However, elevated activities of hepatic enzymes and altered acyl composition of microsomal phosphatidylcholine returned to levels present in controls within four weeks of exclusion of PFOA from the diet (Uy-Yu *et al.* 1990b). Moreover, female rats responded only slightly to the administration of PFOA over short- and long-term exposures, which highlights a marked gender-specific difference in the hepatic response to PFOA.

Mammalian Oral Toxicity – Chronic

In common with sub-chronic animal studies of oral PFOA exposure, studies have also been carried out to determine the effects of PFOA from chronic oral exposure of monkeys, rats, and mice. These chronic exposure studies provided a greater understanding of tumor incidences in mice and rats. Moreover, developmental and reproductive effects were found following a two-generational rat study, and a male-fertility mouse study that included delays in development and increased incidence of neonatal mortality in response to PFOA exposure. Additionally, studies have explored the effects of gestational or lactational exposure to PFOA on mammary gland development and subsequent effects that were observed in the offspring at maturity. The pertinent studies used in this assessment are summarized in Table 2.3.

For example, it was found that chronic exposure to PFOA in mammals frequently led to weight loss. Body weight effects were seen in many studies (Butenhoff *et al.* 2004, 2012, Biegel *et al.* 2001), and present a highly relevant toxicological end point, especially under conditions where there was no decrease in food intake, and yet body weight decreased in neonates (Butenhoff *et al.* 2004). Additionally, in a two-generational study, developmental delays in male and female rats were evident (Butenhoff *et al.* 2012). Studies by Butenhoff *et al.* (2012), and a chronic single dose study by Biegel *et al.* (2001) demonstrated testicular effects in PFOA-exposed animals. Additionally, the Butenhoff *et al.* (2012) two-year bioassay study fed male rats with the equivalent dose of 1.3 or 14.2 mg/kg-d PFOA and female rats with the equivalent dose of 1.6 and 16.1 mg/kg-d PFOA. Exposed animals subsequently failed to exhibit any differences in survival, compared with non-exposed, control animals. This study reported slight increases in ataxia in PFOA-treated female rats toward the end of the study. Body weights were reduced in male rats that received the 14.2 mg/kg-d dose, but effects were not previously seen in female rats until the study had concluded.

At various times throughout the study, statistically significant decreases in the numbers of circulating erythrocytes, hemoglobin levels, and hematocrit percentages were seen in the high-dose group; however, there was no consistent effect

TABLE 2.3
Summary of Chronic Oral Toxicity for PFOA in Mammalian Species

		Test Results			
Test Organism	Test Duration	NOAEL (mg/kg-d)	LOAEL (mg/kg-d)	Effects Observed at the LOAEL	Reference
Rats (male)	2 years	NA	13.6	Reduced body weight gain from reduced food use efficiency, increased Leydig cell hyperplasia and adenomas, increased pancreatic acinar cell proliferation and hyperplasia, increased serum estradiol, luteinizing hormone, follicle stimulating hormone, and prolactin concentrations	Biegel et al. 2001
Crab-eating macaques (male)	6 months	NA	3.0	Increased liver weights, reduced total thyroxin	Butenhoff et al. 2002
Rats	2 years	Male: 1.3 Female: NA	Male: 14.2 Female: 1.6	Increased serum albumen; altered hematology, liver pathology, testicular masses, Leydig cell adenomas in males and increased absolute and relative kidney weights in females	Butenhoff et al. 2012
Rats (female)	2-generation	NA	1.0	Altered weanling induced mammary gland involution and offspring mammary gland development	White et al. 2011b
Rats	2- generation	Male: NA Female: 10	Male: 1.0 Female: 30	Increased absolute and relative liver weights in males; decreased survival in female offspring and delayed sexual maturation	Butenhoff et al. 2004

Legend:
NOAEL: no-observed-adverse-effect level
LOAEL: lowest-observed-adverse-effect level
NA: not applicable

identified in the low-dose group (Butenhoff *et al.* 2012). In addition, hematological end points were unaffected by PFOA treatment in female rats. Throughout the study, there were also statistically significant increases in the mean albumin concentrations in the 14.2 mg PFOA/kg-d male group, relative to the controls, and at three and six months in the 1.3 mg PFOA/kg-d male rat group. There was a statistically significant increase in relative kidney weights (relative to the controls) for PFOA-treated females only. Additionally, lower pituitary weight parameters were found in all of the 14.2 mg/kg-d PFOA-treated male rats, compared with the controls (Butenhoff *et al.* 2012).

Similarly, in the Biegel *et al.* (2001) two-year bioassay study, dietary exposure of male CD rats to an equivalent PFOA dose of 13.6 mg/kg-d subsequently reduced body weight gains in PFOA-treated rats, by a mechanism that was primarily dependent upon reduced food-utilization efficiency. There was also no effect of PFOA exposure on food consumption or survival in PFOA-treated rats (Biegel *et al.* 2001).

In the Butenhoff *et al.* (2002) study, PFOA was administered orally for 6 months to male crab-eating macaques (n = four to six per dose) at doses of 3–30 mg/kg-d. Treatment at the highest dose of PFOA was halted after 12 days, and subsequently reinitiated on day 22 at the reduced dose of 20 mg/kg-d as the highest exposure decreased body weights, due, in part, to reduced food consumption, and decreased THE release of fecal matter.

Monkeys that received 20 mg/kg-d PFOA displayed reduced body weight, with exposure to PFOA causing reduced food consumption. Some of the animals that were dosed at 20 mg/kg-d were no longer administered PFOA due to significant weight loss and reduced food consumption. In addition, one of the animals that was dosed in the partial 30/20 mg/kg-d PFOA group was found in a moribund state and euthanized on day 29, where it was subsequently found that the animal had evidence of dosing injury and lesions of the liver. No weight loss or marked clinical signs were seen in the majority of animals dosed at 3 or 10 mg/kg-d PFOA. However, one of the animals in the low-dose group exhibited hind-limb paralysis and ataxia, failed to respond to painful stimuli, and was not consuming food.

Clinical chemistry analysis revealed increased levels of triglycerides in monkeys that received 20 mg/kg-d PFOA. No other notable findings were found in this high-dose group of animals following whole blood or urological analyses. In addition, no clinical effects were seen in any of the monkeys dosed at 3 or 10 mg/kg-d PFOA, with the exception that one moribund animal was found in the 3 mg/kg-d dosed group. It was suggested that the animal was likely sick prior to commencing this bioassay. All animals in the 20 mg/kg-d dosage group showed increased absolute and relative liver weights. Additionally, none of the PFOA-treated animals showed any gross morphological or histopathological findings (Butenhoff *et al.* 2002).

In a complementary two-year bioassay study (Butenhoff *et al.* 2012), male Sprague Dawley rats were fed the equivalent of 1.3 or 14.2 mg/kg-d PFOA, with female rats being fed 1.6 or 16.1 mg/kg-d PFOA, which was equivalent to the low (30 ppm) and high (300 ppm) doses of PFOA respectively found in the diet. Actual PFOA doses were determined for each 2 week period for both male and female rats in each dosing group and equivalently expressed as mg/kg per day. This dosing regimen also did not affect survival. Ataxia was slightly increased in females toward the end of the study. Body weights were reduced in males that received 14.2 mg/kg-d PFOA; however, no effect was observed in female rats by the completion of the study. Statistically significant decreases in the frequency of erythrocytes, hemoglobin levels, and hematocrit percentages were found at various times in the high-dose group; however, there was no consistent effect found in the low-dose group. PFOA-treated female rats showed no changes in their hematology, compared to the controls.

Throughout the study, statistically significant increases were observed in mean albumin concentrations in male rats that were dosed at 14.2 mg/kg-d PFOA, whereas significant increases were observed in male rats dosed with 1.3 mg/kg-d at three and six months, relative to control animals. There was also a statistically significant increase in relative kidney weight for PFOA-treated female rats only. In all male rats treated with 14.2 mg/kg-d PFOA, it was found that pituitary weight parameters were lower than for the controls (Butenhoff *et al.* 2012).

Mammalian Oral Toxicity – Chronic: Reproduction

Several animal studies have investigated reproductive and fertility outcomes in response to exposure to PFOA, with some studies conducted on rats showing no clear effects on exposure to PFOA (Butenhoff *et al.* 2004a, York *et al.* 2010), whereas others showed that PFOA adversely affected fertility in male mice (Lu *et al.* 2015). Additional evidence indicated decreased body weight and survival in mouse gavage studies at PFOA doses of at least 1 mg/kg-d, and increased incidence of full litter resorptions and stillbirths at doses of at least 5 mg/kg-d for exposures at GD1–17. In addition, exposure to PFOA increased the time to parturition at 10 mg/kg-d, and decreased gains in maternal body weight at doses of 20 mg/kg-d PFOA or higher, all of which were observed for exposures at GD1–17 (Abbott *et al.* 2007, Lau *et al.* 2006, White *et al.* 2007, Wolf *et al.* 2007).

Several other studies have also reported changes in post-natal development in both mice and rats in response to PFOA (Table 2.3). Butenhoff *et al.* (2004a) reported reduced postnatal growth and subsequent delays in development in rats exposed to PFOA. In this two-generational study of dietary exposure to PFOA, reduced gains in body weight prior to weaning and retarded sexual maturity in first-generation male and female rats were seen at 30 mg/kg-d (Butenhoff *et al.* 2004a). Similarly, Yang *et al.* (2009) reported delayed vaginal opening in BALB/c mice at 1 mg/kg-d PFOA or higher, and in C57Bl/6 mice at 5 mg/kg-d or higher, with exposure beginning at PND 21. However, decreased body weights in C57Bl/6 and Balb/c mice were not seen until PFOA doses exceeded 10 mg/kg-d (Yang *et al.* 2009). Wolf *et al.* (2007) conducted cross-fostering mouse studies and showed that gestational PFOA exposure greatly decreased postnatal body weight, delayed both eye opening and body hair growth, and decreased offspring survival.

This research group also found that variation in gestational exposures to PFOA differentially influenced the offspring. For example, under conditions of extended gestational exposures, greater decreases in body weight were found in both male and female mouse pups (Wolf *et al.* 2007). Moreover, in male mice, differences in body weight persisted until PND 92. At a dose of 5 mg/kg-d, offspring that were exposed at GD 7–17 or 10–17 exhibited delayed eye opening and body hair growth; however, decreased post-natal survival was also seen at 5 mg/kg-d when the offspring were exposed at GD 15–17.

In studies by Macon *et al.* (2011) and Tucker *et al.* (2015), maternal exposure to doses of 0.01 mg PFOA/kg or higher delayed mammary gland development in female CD1 mice . When compared with controls, significant differences in the quantitative determination of longitudinal and lateral growth, as well as numbers of terminal end buds, were reported in response to exposure to 1 mg/kg-d PFOA (Macon *et al.* 2011). By contrast, other groups did not find any marked differences in the average length of mammary gland ducts or the average number of terminal end buds per mammary gland per litter in female PPARα wild-type, PPARα-null, or hPPARα sv/129 mouse pups following maternal exposure to 3 mg/kg-d (Albrecht *et al.* 2013). A drawback to these studies was that approaches used to score mammary gland development across the studies were inconsistent, which made comparisons challenging. However, Tucker *et al.* (2015) found that CD-1 mice were markedly more sensitive to the effects of PFOA exposure on mammary gland development (NOAEL = 0.01 mg/kg-d) than their C57Bl/6 counterparts (NOAEL = 0.1 mg/kg-d), when both studies used the previously described scoring approach (Macon *et al.* 2011).

White *et al.* (2011b) conducted a very complex exposure study, that included exposing CD-1 mice to PFOA at doses of 1 and 5 mg/kg-d PFOA *via* oral gavage, which was conducted only during gestation, but also included exposure to drinking water that was supplemented with 5 ppb PFOA that commended during gestation (at GD 7) and continued through to the F_2 generation. Therefore, this study represented a three-generational study. There was no significant effect of PFOA on parental dam gestational weight gain or implant number. PFOA given during gestation at 5 mg/kg-d reduced the number of live fetuses, prenatal survival, and post-natal offspring growth and survival; however, similar effects were not observed at 1 mg/kg-d PFOA or on exposure to PFOA-supplemented drinking-water.

Normal weaning-induced mammary involution was compromised in all PFOA-treated parental dams, including those with only low-dose exposures *via* drinking water. In addition, F_1 offspring exhibited reduced mammary gland scores at PND 22, 42, or 63 in all groups receiving PFOA. Reduced body weights and increased liver weights were transiently observed in the 5 mg/kg-d PFOA-treated dose group, whereas F_1 dams exposed to chronic drinking water concentrations or to gestational PFOA did not exhibit any maternal toxicity. F_1 dams treated with 5 mg/kg-d during gestation exhibited reduced numbers of implants and live fetuses. Unlike F_1 females, developmental mammary gland scores in F_2 females did not differ in association with maternal exposure; however, control F_2 females exhibited unusually low mammary gland scores at PND 10 and PND 22, which made it challenging to determine reliably whether the scores for the F_2 generation were impacted by PFOA exposure (White *et al.* 2011b).

Male and female rats received 1, 3, 10, or 30 mg/kg-d PFOA in the parental and F_1 generations (Butenhoff *et al.* 2004). The parental generation was treated for 70 days prior to cohabitation, and the F_1 was treated from weaning onward. Reproductive performance in the parental generation was unaffected by PFOA treatment in this study. The numbers of pups delivered live or stillborn were also unaffected. There was a slight but significant increase in the number of dead pups in the F_1 following exposure to 30 mg/kg-d PFOA as compared the non-exposed dams. In addition, no reproductive effects in the F_1 generation could be attributed to PFOA treatment. The numbers of pups (F_2) delivered and their survival through lactation were unaffected by PFOA treatment.

In the F_1 generation, mortality of both male and female rats increased in the 30 mg/kg-d groups, with the additional deaths occurring primarily within the first few days post-weaning. Most deaths were likely associated with failure of the pups to thrive within a few days of weaning and were seen almost exclusively in rats that were small at weaning. Statistically significant delays in sexual maturation were seen in F_1 male and female rats in the 30 mg/kg-d dose group, as compared with controls. Low incidences of dehydration (4/30), urine-stained abdominal fur (3/30), and ungroomed fur (3/30) were present in the 30 mg/kg group of parental (P) generation male rats. At the end of the exposure period, body weights were statistically lower in the P generation male rats that had been exposed to PFOA at 3 mg/kg-d or higher. Increased absolute and relative liver weights were also seen in PFOA-treated groups of male rats. Kidney weights, relative to body weight, were also statistically increased in all PFOA-treated groups, and the magnitude of the increase was similar across groups which were administered 3 mg/kg-d PFOA or higher. No treatment-related clinical signs were observed in the P generation female rats at any of the PFOA doses tested (Butenhoff *et al.* 2004).

Statistically significant observations in F_1 male rats were emaciation in the 10 and 30 mg/kg-d dose groups, and urine-stained abdominal fur, decreased motor activity, and abdominal distention in the 30 mg/kg-d dose group. In the 30 mg/kg-d PFOA dose group, statistically significant decreases in body weight and in body weight gain were observed during the juvenile (i.e., through to 35 days of age) and peripubertal (i.e., through to 55–60 days of age) growth phases.

At necropsy, ten areas of discoloration in the liver were present at incidences of 6/60, 10/60, and 9/60 of F_1 male rats that were treated with 3, 10, and 30 mg/kg-d PFOA, respectively. Treatment-related microscopic changes in these livers involved hepatocellular hypertrophy and, less commonly, focal to multifocal hepatocellular necrosis. Hypertrophy and vacuolation of the zona glomerulosa of the adrenal gland was present in 7/10 F_1 male rats that were administered 30 mg/kg-d PFOA. Unlike P generation males, no adrenal changes were seen at a PFOA dose of 10 mg/kg-d. All liver weight parameters (i.e., absolute and relative to both brain and terminal body weight) were increased in all of the PFOA-treated groups in the F_1 generation. In addition, statistically significant increases in kidney weights relative to body weight were seen in all PFOA-treated groups and, concordant with the P generation, the magnitude of this effect was similar across groups that were administered PFOA at 3 mg/kg-d or higher (Butenhoff *et al.* 2004).

Mammalian Oral Toxicity – Chronic: Hormone Effects

Dietary exposure of male rats to PFOA, involving consumption of a concentration that was equivalent to 13.6 mg/kg-d, for 2 years, produced increased levels of serum estradiol, but did not alter testosterone concentrations. The treated group had increased levels of luteinizing hormone, follicle-stimulating hormone and prolactin (Biegel *et al.* 2001). Additionally, oral doses of 3 to 30 mg/kg-d (reduced to 20 mg/kg-d after a few weeks) PFOA, provided for 6 months to male crab-eating macaque monkeys, reduced total thyroxin concentration in all treatment groups. Free thyroxin was reduced in monkeys that had received 10 mg/kg-d or 20 mg/kg-d of PFOA. In addition, levels of testosterone and estradiol were also reported to have decreased in monkeys receiving 20 mg/kg-d (only 2/6 animals) (Butenhoff *et al.* 2002).

Mammalian Oral Toxicity – Chronic: Enzyme Effects

Male and female rats respond differently to PFOA exposure in the diet. Dietary exposure to, ppm PFOA in either male or female rats for 26 weeks increased microsomal 1-acyl-glycerophosphocholine (GPC) acyltransferase and peroxisomal β-oxidation activities in kidneys; however, the observed increase in males was considerably greater than that in females (Kawashima *et al.* 1991). Exposure of male and female rats to a lower dietary PFOA concentration of 100 ppm for 26 weeks caused increased liver peroxisomal β-oxidation and liver catalase levels in male and female rats; however, decreases in liver glutathione (GSH) peroxidase were seen only in male rats. Additionally, liver cytosolic GSH S-transferase activity had decreased in both male and female rats, although the levels of this enzyme changed comparatively less so in female rats (Kawashima *et al.* 1991). Activities of liver 7-ethoxycoumarin *O*-deethylase, aminopyrine *N*-demethylase, cytochrome P-450, and aniline *p*-hydroxylase were all increased in male rats but remained unaltered in females (Kawashima *et al.* 1994). Relevant studies used in this assessment are summarized in Table 2.3.

Male rats, that were fed diets with PFOA at doses equivalent to 1.3 and 14.2 mg/kg-d for two years, exhibited clinical biochemistry findings at three months. In particular, slight increases were observed in the mean activities of alanine aminotransferase (ALT), aspartate aminotransferase (AST), and alkaline phosphatase (ALP), and a moderate decrease in mean creatine phosphokinase (CPK) activity was observed in male rats that had received either doses of PFOA in the diet. From 6 to 18 months, mean ALT, AST, and ALP activities for the PFOA-treated males were higher than the control values, with some of these values statistically significantly higher in both treatment groups than the control values. By contrast, at 24 months, activities of those enzymes were higher only in the group receiving a dietary PFOA dose of 14.2 mg/kg-d. Dietary exposure of female rats to 1.6 or 16.1 mg/kg-d promoted no significant effects (Butenhoff *et al.* 2012). A summary of the oral ingestion health effects of PFOA in the class Mammalia is presented in Figure 2.1.

MAMMALIAN INHALATION TOXICITY

Inhalation toxicity data in laboratory animals are limited in scope to acute single or repeated exposures, that were designed for pharmacokinetic studies. Only one developmental toxicity study on PFOA was found in the literature, and this was conducted on rats. Sub-chronic or chronic inhalation toxicity studies on mammals had not been published for assessment at the time this review was researched. In general, the observed adverse effects of an inhalational exposure to PFOA were similar to those effects observed after exposure to an irritating dust following inhalation exposure to PFOA. For male rats exposed to PFOA as a dust in the air, the 4-hour LC_{50} (lethal concentration 50%) was 980 mg/m^3, with adverse clinical signs of body weight loss, irregular breathing, red discharge around the nose and eyes, and corneal opacity and

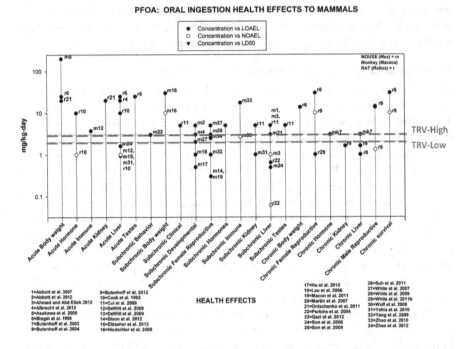

FIGURE 2.1 Oral ingestion health effects for mammalian species: TRVs at the high and low derivations were determined by benchmark dose analyses (BMD/BMDL$_{10}$) (after DeWitt *et al.* 2008).

corrosion (Kennedy *et al.* 1986, 2004). Additionally, there were no published reports of any tumorigenic effects of chronic inhalational or dermal exposures to PFOA in the current literature.

MAMMALIAN INHALATION TOXICITY – ACUTE

Exposure of male rats to PFOA for 4 hours to air concentrations that ranged from 380 to 5,700 mg/m³ caused mortality at all exposure levels (Kennedy *et al.* 1986). For male rats, the acute (4-hour) LC$_{50}$ was 980 mg/m³ but a NOAEL could not be established. Rats exposed to 810 mg/m³ or more showed corneal opacity and corrosion, which was confirmed by fluorescein staining. During exposure, rats displayed clinical signs of body weight loss, gasping, irregular breathing and a red discharge around the eyes and nose. Surviving rats lost weight for an initial 1–2 days after exposure but resumed normal weight gain thereafter. Microscopic examination showed acute pulmonary edema in the exposed rats (Kennedy *et al.* 1986).

MAMMALIAN INHALATION TOXICITY – SUB-CHRONIC

Exposure of male rats to PFOA for 6 hours in an exposure cycle of 5 days on and 2 days off, with an additional 5 days of exposure to air concentrations of 1, 8, or 84

mg/m^3 caused light to mild nasal and ocular discharge. These clinical signs were more greatest in rats that were exposed at the higher dose of 84 mg/m^3, and somewhat elevated in animals exposed to PFOA at 8 mg/m^3 when compared to controls or other exposed groups (Kennedy *et al.* 1986). Animal mortalities occurred only at a dose of 84 mg/m^3. No adverse body weight effects were observed in the two lower-concentration test groups; however, the body weights of rats in the 8 mg/m^3 dosed group were higher than those found in the control group.

Serum alkaline phosphatase activity was elevated immediately after exposure of rats to 8 or 84 mg/m^3 PFOA, and remained at an increased inhalation level in the 84 mg/m^3 group following a 14-day recovery period. Absolute and relative liver weights were elevated in rats that were exposed to 8 or 84 mg/m^3, but not to the lowest dose of 1 mg/m^3. Repeated inhalation exposure of rats to PFOA caused panlobular and centrilobular hepatocellular hypertrophy and necrosis of the liver (Kennedy *et al.* 1986).

In similar studies, female rats (at n = 6 dams per group) were exposed to PFOA dust concentrations at 0, 0.1, 1, 10, or 25 mg/m^3 *via* inhalation for 6 hours per day at GD 6–15 (Staples *et al.* 1984). Three females that were exposed to 25 mg/m^3 died. Food consumption was reduced in female rats that had been exposed to 10 or 25 mg/m^3.

Dams that were exposed to PFOA at 25 mg/m^3 showed reduced body weight. Liver weight in dams that were exposed to 25 mg/m^3 was increased at GD21, and relative liver weight was increased in rats exposed to 10 or 25 mg/m^3. No effect was seen in terms of maintaining pregnancy or in terms of the incidence of resorptions at any exposure level, compared with the control rats. The mean fetal body weight in the 25 mg/m^3 PFOA group was also decreased (Staples *et al.* 1984).

Finally, other research groups studied the toxicokinetics of PFOA dust exposures, where unique toxicokinetic differences between male and female rats were found following exposure to single or repeated inhalation exposures (Hinderliter 2003). Sprague Dawley rats were exposed nose-only to PFOA aerosols at concentrations of 0, 1, 10, or 25 mg/m^3 for 6 hours or in complementary studies for 6 h/d, 5 d/week, for three weeks (Hinderliter 2003). Absorption was indicated in both male and female rats after single and repeated exposures. A summary of the sub-chronic inhalation toxicity data of importance to TRV development for PFOA is presented in Table 2.4.

Plasma PFOA concentrations were found to be proportional to the exposure concentration. The maximum serum concentration (C_{max}) values was approximately 2- to 3-fold higher in male than in female rats. In addition, the C_{max} values were sustained for up to 6 hours in male rats as compared with just 1 hour in females. Similarly, female rats eliminated PFOA rapidly at all exposure levels. In addition, the plasma levels of PFOA had dropped below the analytical Limit of Quantification (LOQ) (0.1 µg/mL) by 12 hours after exposure in female rats.

In male rats, the plasma concentration remained at approximately 90% of the C_{max} concentration at all exposure levels for 24 hours post-exposure. In addition, the steady-state level was reached following repeated exposures. Although these results clearly show toxicokinetic differences between male and female rats, toxicity data were not included, limiting the use of this information in a quantitative

Risk to Wildlife from Exposures to PFAS

TABLE 2.4
Summary of Sub-Chronic Inhalation Toxicity for PFOA in Mammalian Species

Test Organism	Test Duration/ Period	NOAEL (mg/ kg-d)	LOAEL (mg/ kg-d)	Effects Observed at the LOAEL	Reference
Rats (male)	2 weeks	1	8	Elevated serum alkaline phosphatase activity, liver weights, and liver histopathology	Kennedy *et al.* 1986
Rats (female)	GD6–15	1	10	Clinical signs of wet abdomens, blood-stained tears and nasal discharge, reduced food consumption; increased relative liver weights	Staples *et al.* 1984

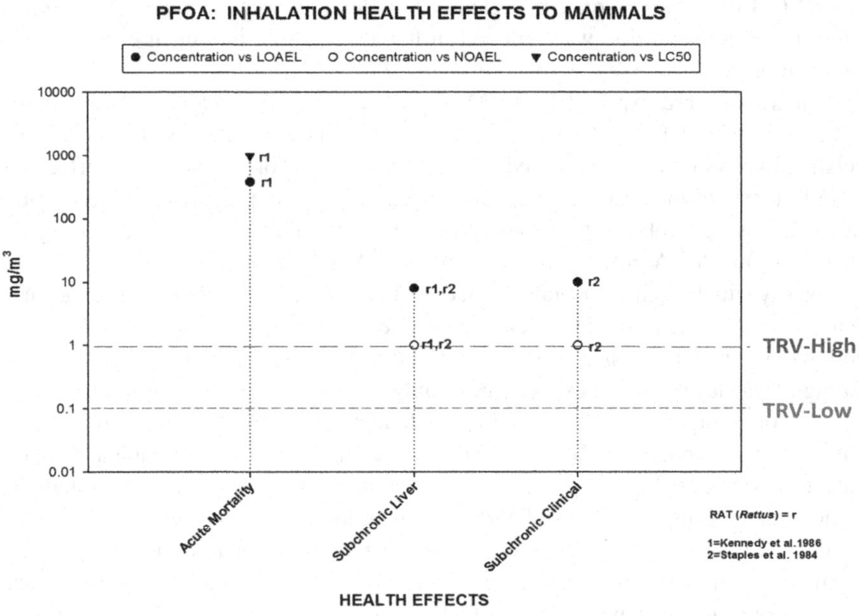

FIGURE 2.2   Inhalation health effects for mammalian species.

risk assessment (Hinderliter 2003). A summary of the inhalational health effects of PFOA in the class Mammalia is presented in Figure 2.2.

DERMAL TOXICITY

There are very early studies from the 1980s, which demonstrate dermal absorption following exposure to PFOA from animal studies conducted on rabbits and rats

TABLE 2.5
Summary of Acute Dermal Toxicity for PFOA in Mammalian Species

Test Organism	LD$_{50}$ (mg/kg)	NOAEL (mg/kg-d)	LOAEL (mg/kg-d)	Effects Observed at the LOAEL	Reference
		Test Results			
Rabbits (male)	4,300	NA	1,500	Labored breathing, weight loss, and skin irritation, with formation of a large crusty area at the application site	Kennedy 1985
Rats (male and female)	Male: 7,000 Female: >7,500	Male: 5,000 Female: 5,000	Male: 7,500 Female: 7,500	Reduced body weight	Kennedy 1985
Mice (female)	NA	2.5	6.25	Increased liver weights	Fairley et al. 2007

(O'Malley and Ebbins 1981, Kennedy 1985). Groups of four New Zealand white rabbits per dose (n = 2 male and n = 2 female) were dermally exposed to 100, 1,000, or 2,000 mg/kg PFOA for 14 days. All animals failed to survive dermal exposure to 2,000 mg/kg PFOA, with three of the four animals in the 1,000 mg/kg dose dying, but with no losses reported at 100 mg/kg (O'Malley and Ebbens 1981). In the study by Kennedy (1985), rabbits and rats were dermally exposed to ten separate applications of PFOA at doses of 0, 20, 200, or 2,000 mg/kg. Following treatment, heightened dose-dependent levels of blood organofluorine (an indirect measure of serum or plasma PFOA, given the technology at that time) were found in both rabbits and rats. These and other relevant studies are summarized and discussed in detail in Table 2.5.

DERMAL TOXICITY – ACUTE

Exposure of the dorsal surface of both ears to up to 50 mg/kg PFOA in female BALB/c mice for four days caused no mortality or inflammation at the site of exposure (Fairley et al. 2007). In addition, no changes were seen in body weight. Animals were sacrificed six days later, at which time there was a significant increase in liver weight caused by exposure to PFOA at 6.25 mg/kg or higher, as compared with control animals. Significant decreases were seen in thymus weight at exposures of up to 50 mg/kg and decreases in spleen weight at exposures of 25 and 50 mg/kg PFOA (Fairley et al. 2007). Thus, the LOAEL was determined to be 6.25 mg/kg PFOA, based on increased liver weight, and the NOAEL was determined to be 2.25 mg/kg PFOA (Fairley et al. 2007). Furthermore, in an in vitro test, PFOA was classified as corrosive, with fewer than 50% of cells being viable after three minutes exposure, and less than 15% remaining viable after 1-hour exposure to 50 mg of concentrated, solid PFOA wetted with water (Franko *et al.* 2012).

Several research groups have determined that the dermal LD_{50} in New Zealand White rabbits and in rat models exceeded 2,000 mg/kg (Kennedy 1985, Glaza 1995). As briefly described above, Kennedy (1985) treated male rabbits (n = 6 per group) with a dermal dose of 1,500, 3,000, 5,000, or 7,500 mg/kg PFOA, which was applied as an aqueous paste. Kennedy (1985) calculated an LD_{50} of 4,300 mg/kg for dermal exposure of rabbits, 7,000 mg/kg for male rats, and 7,500 mg/kg for female rats. Dermal exposure of rabbits to PFOA at 1,500 mg/kg showed evidence of labored breathing, weight loss, and skin irritation, with the formation of a large crusty area at the application site. In addition to these observations, rabbits treated with 3,000 mg/kg were somewhat lethargic and a single death from a group of n = 6 treated animals, occurred 7 days after treatment. At 5,000 mg/kg, deaths occurred within 3 to 4 days (Kennedy *et al.* 2004). Clinical signs included nasal discharge, pallor, diarrhea, weakness, wet staining of the underside of the body, severe weight loss, and severe skin irritation, along with areas of necrosis. The two rabbits treated with 7,500 mg/kg PFOA died within 3 days and displayed all of the clinical signs previously described, with the exception that, in this case, the local skin reaction did not advance to necrosis but only to moderate redness with slightly raised areas (Kennedy *et al.* 2004). A NOAEL for mortality was established as 1,500 mg/kg, but a NOAEL could not be established for clinical signs (Kennedy 1985).

In related work (Kennedy 1985), rats were treated dermally with 3,000, 5,000, or 7,500 mg/kg PFOA as an aqueous paste. The LD_{50} was 7,000 mg/kg and >7,500 mg/kg in male and female rats, respectively. Rats treated with 3,000 mg/kg had only a mild, transient weight loss, in which approximately 5 to 7% of their initial body weight was lost for one to four days post-treatment. This was followed by resumption of normal weight gain, with no local signs of skin irritation being seen. Male and female rats treated at 5,000 mg/kg both showed mild skin irritation and a body weight loss that was mild to moderate (10% of the initial body weight was lost). Male rats were lethargic and had some staining of the pelvic area. Deaths seen at 7,500 mg/kg PFOA occurred within 3 days of treatment. Lethargy was more pronounced in males, and staining of the face and pelvic area was more apparent at the higher doses. The NOAEL was 5,000 mg/kg for mortality for both male and female rats, but no NOAEL could be established for clinical signs in either sex (Kennedy 1985).

As mentioned above, male rats were treated dermally with 20, 200, or 2,000 mg/kg PFOA for 5 days, not treated for 2 days, then treated for another 5 days (Kennedy *et al.* 2004). No mortality occurred in any group. Some evidence of skin irritation was seen at the dose application site of rats treated with either 200 or 2,000 mg/kg, but not when exposed to 20 mg/kg. Rats treated with 2,000 mg/kg PFOA salivated sporadically throughout the exposure period and had wet staining of the pelvic area during the first week of a 42-day recovery period. Red blood cell (RBC) counts decreased, relative to the control, in the 20 mg/kg group after 10 days exposure and after 14 days of a 42-day total recovery period. After 10 days exposure, it was also found that liver weights and relative liver weights were reduced in animals exposed to 20 mg/kg PFOA. The NOAEL for mortality was 2,000 mg/kg, and the NOAEL based on both body weight and clinical signs was 200 mg/kg. A NOAEL was not derived for RBC counts or liver weights (Kennedy 1985). In a series of related studies (Gabriel 1976b, 1976c), PFOA was shown to be an ocular irritant in rabbits under conditions where it was not washed from the site of exposure, but lacked irritancy if the compound was washed away immediately from the eye. Earlier controversial studies also presented conflicting reports on the skin irritancy of PFOA. In one study, PFOA was found to be a skin irritant in rabbits (Markoe 1983); however, a study by Gabriel (1976d) reported that PFOA was not a skin irritant.

Dermal Toxicity – Acute: Immunotoxicity

The role of experimental dermal exposure to PFOA in promoting any toxicological effect was studied in BALB/c mice (Fairley *et al.* 2007). Animals were exposed to PFOA at doses of 0, 0.25, 2.5, 6.25, 12.5, 25, or 50 mg/kg-d by application to the dorsal surface of both ears for four days, following which mice were sacrificed on day 10. Dermal exposure to PFOA did not decrease body weight of the animals or cause any signs of inflammation at the dermal site of exposure. However, in animals that were dosed with 6.25 mg/kg-d PFOA, liver weights increased as compared with control liver weights ($P < 0.01$). From this study, the LOAEL was estimated to be 6.25 mg/kg-d, which was based on increased liver weights, and the NOAEL was estimated at 2.25 mg/kg-d (Fairley *et al.* 2007).

DERMAL TOXICITY – SUB-CHRONIC

There are currently no published data on the effects of sub-chronic dermal exposure to PFOA (Figure 2.3).

MAMMALIAN TOXICITY – OTHER

Mammalian Toxicity – Other: Endocrine Disruption

Further studies are needed to identify the novel pathways that modulate the endocrine-disrupting effects of PFOA, in addition to the doses at which they are active and physiologically relevant (White *et al.* 2011a). In an earlier study (Thottassery *et al.* 1992), the role played by adrenal hormones on liver enlargement and peroxisomal proliferation was explored following exposure of male adrenal-deficient and adrenal-gland intact Sprague Dawley (SD) rats to a single dose of 150 mg/kg PFOA. SD rats lacking adrenal glands were dosed two days after surgery and all animals were sacrificed two days after PFOA exposure.

Exposure to PFOA increased whole liver peroxisomal β-oxidation activity levels similarly among the PFOA-treated groups of rats with or without an intact adrenal gland (Thottassery *et al.* 1992). Moreover, in intact rats, but not in rats missing the adrenal cortex, exposure to PFOA increased whole liver catalase activity. The experimental observations suggested that adrenal hormones were not essential for inducing peroxisomal β-oxidation activity in rats exposed to PFOA.

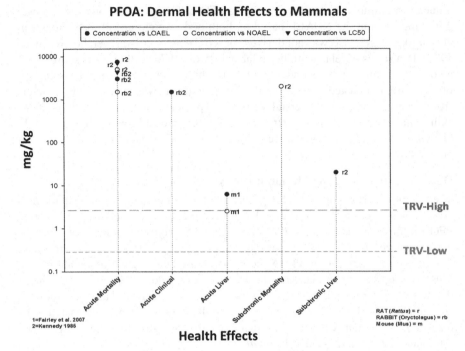

FIGURE 2.3 Dermatological health effects of the class Mammalia.

Mammalian Toxicity – Other: Genotoxicity/Mutagenicity

A number of *in vivo* and *in vitro* assays were employed to determine the genotoxicity of PFOA. In an early mammalian (mouse embryo fibroblasts) cell transformation and cytotoxicity assay, that measured both colony and focal transformation, there was no evidence of a dose-dependent effect on treatment with PFOA (Garry and Nelson 1981). PFOA was also studied for its ability to induce mutations in the *Salmonella typhimurium–Escherichia coli*/mammalian-microsome reverse mutation assays (Lawlor 1995, 1996). Another mutagenicity study (Lawlor 1995, 1996) in *S. typhimurium* gave a positive result; however, the result could not be reproduced. The analysis showed that, irrespective of whether or not metabolic activation was employed in the assays, PFOA failed to induce mutations in either *S. typhimurium* or *E. coli*. Additionally, PFOA failed to induce chromosomal aberrations in human T cells when assayed with/without metabolic activation (Murli 1996a). In other clastogenic studies (Murli 1996b, 1996c), PFOA was tested for its induction of chromosomal aberrations in Chinese hamster ovary (CHO) cells. In one study (Murli 1996b), PFOA induced both chromosomal aberrations and polyploidy with/without metabolic activation. In the subsequent study, however, significant increases in chromosomal aberrations in the absence of metabolic activation were not seen (Murli 1996c). These effects were seen only at toxic concentrations of PFOA (EFSA 2008).

Other researchers showed that PFOA, at concentrations of up to 500 μM, was not mutagenic in the absence or presence of metabolic activation by liver S9 fractions in *S. typhimurium* strains TA98, TA100, TA102, or TA104 (Fernández Freire *et al.* 2008). Furthermore, although PFOA is likely to be non-mutagenic (USEPA 2002), other groups have reported that the filamentous network of mitochondria was fragmented after 24 hours exposure to PFOA in Vero cells, which is a cell-line derived from monkey kidneys (Fernández Freire *et al.* 2008). PFOA caused increased production of hydrogen peroxide (a reactive oxygen species) in Vero cells at 500 μM. Exposure to PFOA resulted in apoptotic Vero cells at concentrations greater than 50 μM (Fernández Freire *et al.* 2008).

Syrian hamster embryo (SHE) cells were exposed to PFOA in culture to determine whether PFOA was cytotoxic or genotoxic. The cytotoxicity assay of PFOA tested a range of concentrations from 3.7×10^{-5} to 1,390 μM, showing no cytotoxicity up to 186 μM, but a 100 percent decrease in plating efficiency at the two highest concentrations tested (557 and 1,390 μM). In the two-stage (initiation–promotion) cell transformation assay protocol, PFOA was tested in the range of concentrations from 3.7×10^{-4} to 37 μM in two series of experiments conducted in parallel on SHE cells, non-pretreated or pretreated with BaP (0.4 μM) for 24 hours. No cytotoxicity was registered, regardless of treatment. The results of the comet assay applied to PFOA-treated cells revealed no genotoxic effect after 5 hours and 24 hours of exposure to PFOA in the concentration range from 4×10^{-5} to 300 μM (Jacquet *et al.* 2012).

In terms of mutagenic activities, very few published studies were found which investigated the mutagenic properties of PFOA. However, in two studies, PFOA was not found to induce any marked increases in micronucleus frequency and it was concluded not to be a mutagen in the mouse micronucleus assay (Murli 1995, 1996d).

In summary, the assay systems described, including assays conducted with human lymphocytes (Murli 1996c, NOTOX 2000), CHO cells (Murli 1996b, 1996c, Sadhu 2002), SHE cells (Jacquet *et al.* 2012) and *S. typhimurium* (Lawlor 1995, 1996, Friere *et al.* 2008) indicated that PFOA is unlikely to be genotoxic or mutagenic.

Mammalian Toxicity – Other: Effects on Enzymes

In comprehensive studies (Liu *et al.* 1996), adult male Crl:CD BR (CD) rats (n = 15/group) were treated with a dosage range of PFOA concentrations by oral gavage for 14 days in an attempt to identify any modulation of aromatase activity. It was found that PFOA treatment did not affect testicular aromatase activity. However, in rats treated with 2 mg/kg-d PFOA, when adjusting for liver weight and body weight, it was found that the protein yield of isolated hepatic microsomes was markedly increased at 0.2 mg/kg-d PFOA, with hepatic aromatase activity, total hepatic aromatase activity, and serum estradiol concentration all increased on exposure to 2.0 mg/kg-d PFOA ($P < 0.05$). Additionally, total hepatic aromatase activity and serum estradiol concentration were highly positively correlated ($P < 0.001$).

In the above study (Liu *et al.* 1996), that determined the effects of PFOA on aromatase activity, the effects of PFOA on peroxisome β-oxidation and cytochrome P450 activities were also explored. Male CD rats (n = 15/group) were orally exposed to PFOA at 0, 0 pair-fed, 0.2, 2.0, 20, or 40 mg/kg-d for 14 days. From isolated liver samples, total cytochrome P450 levels were increased ($P < 0.05$) at doses of 20 mg/

kg-d PFOA or greater, and the activity of peroxisome β-oxidation was increased at doses of 2 mg/kg-d PFOA or greater. The LOAEL was estimated at 2 mg/kg, based on increased liver weight, serum estradiol concentration, and hepatic aromatase activity, and the NOAEL was estimated at 0.2 mg/kg (Liu *et al.* 1996).

An intraperitoneal dose of 40 mg/kg in mice increased mRNA expression of cytochrome P450 (Cyp) 2B10 (18-fold increase), 3A11 (1.8-fold), and 4A14 (32.4-fold), but did not alter mRNA expression of Cyp1A1/2. Two days after administration, exposure to PFOA increased protein levels of Cyp2B10 and 4A by about 68- and 7-fold, respectively. PFOA also increased nuclear constitutive androstane receptor (CAR) protein levels 4-fold and CAR mRNA expression 1.9-fold. PFOA tended to increase the abundance of peroxisome proliferator–activated receptor-alpha (PPARα) mRNA, but the increase was less than 50 percent and not statistically significant (Cheng and Klaassen 2008).

Mammalian Toxicity – Other: Immunotoxicity

PFOA is immunotoxic in mice (USEPA 2002). Fair *et al.* (2013) reported on the health of Atlantic bottlenose dolphins (*Tursiops truncatus*) sampled from the estuarine coastal areas of Charleston, SC, USA. For immune function, PFOA was shown to function in a PPARα–dependent manner. PFOA was associated with increases in various hematological parameters, such as anion gap function, serum phosphorus concentration, amylase activity, total protein concentration, and total globulin concentration, and decreases in creatine phosphokinase activity. The increased serum phosphorus levels was particularly notable because progressive renal insufficiency leads to hyperphosphatemia, with increased creatinine concentration, coupled with heightened phosphorus levels, being indicative of a prolonged and sustained defect in kidney function and possibly kidney disease.

Although none of these relationships establishes causality, they do provide the first line of evidence that PFCs, including PFOA, may be involved in modulating immunological, hematopoietic, and hepatic-renal function in wild dolphins with high PFC contaminant burdens. Not all findings are likely to be directly related to PFOA exposure but are instead associated with exposure to PFCs in general. Furthermore, compared with non-diseased southern sea otters (*Enhydra lutris nereis*), that were collected from the California coast, concentrations of PFOA were higher in the livers of their diseased counterparts due to the presentation of an infectious disease or from appearing emaciated (Kannan *et al.* 2006). Other contaminants were also present in the livers of southern sea otters, and consequently, no direct causal link can be made between effects reported and PFOA exposure (Kannan *et al.* 2006).

Mammalian Toxicity – Other: Estrogenic Response

MCF-7 BOS breast cancer cells were cultured with PFOA at concentrations of up to 30 μg/mL. Significant cytotoxicity (i.e., based on cell viability) occurred at a dose of 30 μg/mL. PFOA also produced a proliferative response in MCF-7 BOS cells (Henry and Fair 2013). The proliferative response caused by PFOA was 68–95% of the maximal response caused by estrogen – a result that suggested that PFOA might be an agonist of the estrogen receptor. When co-treated with PFOA and 17-β estradiol (or E2), the proliferative response was 21–28% of the response obtained with E2 alone. Thus, from this limited study, it appears that PFOA might be a potent xenoestrogen with intrinsic anti-estrogenic qualities, irrespective of the dose (Henry and Fair 2013).

SUMMARY OF AVIAN TOXICOLOGY

There are few studies reporting the possible impacts of PFOA on birds, most of which involved exposing chicken eggs to PFOA (Smits and Nain 2013, Jiang et al. 2012, Strömqvist et al. 2012, O'Brien et al. 2009, Yanai et al. 2008, Pinkas et al. 2010). These studies indicated that exposure to PFOA can adversely affect developing bird embryos and hatchlings. However, in the context of the aforementioned studies, it is not possible to extrapolate to impacts of PFOA exposure on wildlife, due to an absence of exposure data and concentrations from dosed adults.

The study by Smits and Nain (2013) evaluated the effects of PFOA on 4-week-old male Japanese quail, with a particular focus on immunotoxicological end points. Quail were randomly allocated into three treatment groups of 18 birds per treatment (drinking water alone, PFOA in drinking water at concentrations of 1 and 10 ppm). The quail were exposed for 8 weeks, from 4 weeks to 12 weeks of age at the conclusion of the study. The mean daily PFOA intake per bird was calculated to be 0.2 and 2.1 mg/kg in the 1 and 10 ppm exposure groups, respectively. These doses were similar to those described in the study (described below) by Yeung et al. (2009). After 6 weeks' exposure, quail were challenged with an infectious dose of E. coli to assay for disease resistance or susceptibility as a measure of impact on immunotoxicity following exposure to PFOA (Smits and Nain 2013).

It was also found that, whereas antibody-mediated and innate immune responsiveness were largely unaffected at any dose of PFOA exposure, T cell-mediated immune responses were dampened following exposure to 10 ppm PFOA. However, there was no evidence of any increased or altered morbidity or mortality in the disease responsiveness between the control and PFOA-exposed quail in this study. Thus, although PFOA provoked T cell immunosuppression, this did not enhance disease susceptibility of the quail to E. coli (Smits and Nain 2013).

In another study, fertile chicken (*Gallus gallus*) eggs had 0.5, 1, or 2 mg/kg PFOA injected into the air sac. On day 19 of incubation, embryo mortality in the 2 mg/kg dose group was statistically higher, relative to the vehicle control group (Jiang et al. 2012). In hatchling chicks, relative liver weight was statistically higher in the 0.5 mg/kg and 1 mg/kg dose groups, relative to the vehicle control group. Heart morphology at 1 and 2 mg/kg showed reduced ventricular wall thickness and thinning of the myosin dense layer (Jiang et al. 2012). Heartbeat rate was decreased when eggs were dosed with 0.5 mg/kg and increased when eggs were dosed with 1 or 2 mg/kg. Stroke volume was higher in the 0.5 mg/kg group but was lower in both the 1 and 2 mg/kg groups (Jiang et al. 2012). When 20 µg PFOA/g egg was injected into the air cells of chicken eggs, mortality increased in PFOA-treated eggs (Strömqvist et al. 2012). However, studies of mRNA transcription patterns did not support the viewpoint that embryonic mortality was related to PPARα activation (Strömqvist et al. 2012).

In the study by Yeung *et al.* (2009), one-day-old male chicks were exposed to PFOA by oral gavage at two doses that were defined as high (1.0 mg/kg BW) or low dose (0.01 mg/kg BW), with a saline/ethanol vehicle control, three times per week for three weeks. Following the three weeks of exposure, half of the chicks were sacrificed and the remaining 50% were allowed to depurate for an additional three weeks. No dose-dependent statistically significant differences in body/organ weights were observed among the treatment and control groups after three weeks of exposure or after three weeks of depuration (Yeung *et al.* 2009). In the context of PFOA accumulation in the blood, the half-lives for PFOA at the 0.1 and 1.0 mg/kg doses were both equivalent to 3.9 days. In this study, the liver was confirmed to be the key target of PFOA exposure, and the blood was the predominant reservoir during depuration (Yeung *et al.* 2009).

In a related study, chicken eggs (4 groups at n = 20 eggs/group) were injected with PFOA across the dose range of 10 ng/g egg, 100 ng/g egg, 1 μg/g, and 10 μg/g egg. In addition, 20 eggs each were assigned to an untreated control group, and a DMSO carrier vehicle alone control group (O'Brien *et al.* 2009). The injection volume was 1 μl/g egg as determined from the mean weight of the eggs. Doses had no effect on embryonic survival, gross abnormalities, or relative liver weights. In addition, liver concentrations indicated penetration and assimilation of PFOA, although there was no effect on liver mRNA expression (O'Brien *et al.* 2009). In a related study, fertile chicken eggs received injections of 5, 20, or 40 mg/kg PFOA into the pointed end of the egg (Yanai *et al.* 2008). A high incidence of embryonic death and impaired hatchability at all doses of PFOA was noted. In addition, PFOA-exposed offspring showed two distinct physical anomalies: 1) a high prevalence of splayed legs, and 2) a marked number of chicks showing partially or completely missing typical yellow pigmentation (Yanai *et al.* 2008). Experiments conducted by others (Pinkas *et al.* 2010) injected fertile chicken eggs with 5 or 10 mg PFOA/kg egg into the albumen. Either dose of PFOA decreased embryonic survival and hatching; however, there was no difference seen in terms of body weight when comparing controls and treated hatchlings at the time of hatching. No differences in morphology were seen at the time of hatching. However, both male and female chicks exhibited reduced imprinting scores (Pinkas *et al.* 2010).

In a more recent study that explored the developmental cardiotoxicity of PFOA in birds (Jiang *et al.* 2016), fertile chicken eggs were exposed to 0.1, 0.5, 1.0, 2.0, or 5.0 mg/kg PFOA *via* air sac injection. Exposure to PFOA induced developmental cardiotoxicity in the chick embryo. Markedly increased activities of carnitine acetyltransferase were observed in PFOA-exposed embryonic day six (ED6) chick embryo hearts by Western immunoblotting (Jiang *et al.* 2016). In addition, PFOA-induced developmental cardiotoxicity in chick embryos was associated with decreased concentrations of the cardioprotective amino acid derivatives L-carnitine, acetyl-L-carnitine, and propionyl-L-carnitine levels in ED6 chick embryo hearts (Jiang *et al.* 2016).

Differential sensitivity to the possible developmental toxicity from PFOA exposure was identified following *in ovo* exposure to PFOS, compared with PFOA, in the great cormorant (*Phalacrocorax carbo sinensis*), herring gull (*Larus argentatus*) and the domestic White Leghorn chicken (*G. gallus domesticus*) by direct injection of the test chemical into the air cell (Norden *et al.* 2016). The most sensitive species was the chicken, with a 50 percent reduction in embryonic survival at a PFOS dose of 8.5 µg/g egg, and a PFOA dose of 2.5 µg/g egg. The cormorant was the least sensitive species to PFOA exposure. In addition, the difference in sensitivity to PFASs between the chicken and herring gull was by a factor of 2.7 for PFOS, increasing to 3.5 for PFOA. Between the chicken and the great cormorant, the difference in sensitivity was by a factor of 2.6 for PFOS and 8.2 for PFOA (Norden *et al.* 2016).

This study also revealed that the effects on embryo survival were seen at egg injection doses of PFOS that were similar to the concentrations detected from environmental samples of wild birds, an observation that suggests that PFOS might be influencing highly exposed populations of birds. The study also revealed differential sensitivity of the chicken, herring gull and great cormorant to PFOS and PFOA – a finding that has relevance when determining avian wildlife risk assessment.

In addition to the above studies, increasing interest in avian wildlife responsiveness to PFOA exposure is evident from the large number of recent studies (Groffen *et al.* 2017, Miller *et al.* 2015, Route *et al.* 2014, Vicente *et al.* 2015, Jaspers *et al.* 2013, Braune and Letcher 2013).

In the study by Groffen *et al.* (2017), the presence of perfluoroalkylated acids (PFAAs) in the terrestrial environment was assessed – a departure from the usual route of exploring PFAAs in marine and aquatic environments. Among the PFAAs, PFOA concentrations were determined in the eggs of great tits (*Parus major*) that were collected at a fluorochemical plant and in three other areas at defined distances, between 1 and 70 km (to assess concentration gradient effects), from the source of PFOA pollution in Antwerp, Belgium. The median concentration of 19.8 ng/g for PFOA was among the highest ever reported in bird eggs (Groffen *et al.* 2017). The authors concluded that the levels of PFOA (and other PFAAs determined in this study) were found at levels expected to induce toxicity in avian species.

A comprehensive analysis by Miller *et al.* (2015) linked recent trends in the voluntary phase-outs and restrictions in the use of PFCAs, like PFOS and PFOA, by the international community with trends seen in the concentrations of these compounds in aquatic birds that breed on the Pacific coast of Canada over a period of 20–30 years. This research group studied the following avian species to investigate temporal changes in PFAS levels with dietary changes: ancient murrelet (*Synthliboramphus antiquus*), Leach's storm-petrel (*Oceanodroma leucorhoa*), rhinoceros auklet (*Cerorhinca monocerata*), double-crested cormorants (*Phalacrocorax auritus*), and great blue herons (*Ardea herodias*) (Miller *et al.* 2015).

It was found that the concentration of PFOS, the dominant PFSA compound detected in all five avian species, increased and then decreased in the auklet and cormorant eggs, an observation that was in agreement with the manufacturing phase-out of PFOS and PFOA. However, levels of both PFOS and PFOA continued to increase in petrel eggs and remained largely unchanged in heron eggs. It should be noted that any temporal analysis excluded murrelets, because only data from 2009 were available for this species. The dominant PFAS compounds varied between the offshore (further from land) and coastal (closer to land) species, with increases seen in the offshore species and little or variable changes seen in the coastal species Miller *et al.* 2015). This research group also found that there were minimal temporal changes seen in stable isotope values for $\delta^{15}N$ and $\delta^{13}C$ from samples collected from these avian species from the Pacific coast of Canada. This observation indicated that diet alone was not driving the observed PFAS concentrations (Miller *et al.* 2015).

In a very similar study from the National Wildlife Research Center at Environment Canada (Braune and Letcher, 2013), the levels of PFSAs and perfluorinated carboxylates (PFCAs) were measured in the eggs of a variety of avian species. These included eggs from the thick-billed murres (*Uria lomvia*) and northern fulmars (*Fulmarus glacialis*) from Prince Leopold Island in the Canadian Arctic between 1975 and 2011 and eggs of three additional species (namely the black guillemot (*Cepphus grille*), the black-legged kittiwake (*Rissa tridactyla*), and the glaucous gull (*Larus hyperboreus*), that were sampled in 2008. Interestingly, the authors concluded that, based on already published toxicity thresholds, the levels of PFOA and PFOS found in any of the seabird species from the Canadian Arctic listed above were of toxicological concern.

In a related study, blood specimens from 261 bald eagle nestlings across six study areas from the upper midwestern US area were collected from 2006 to 2011 (Route *et al.* 2014). Differential analysis of 16 PFCs revealed that the sum of all PFCs reached levels as high as 7,370 ng/mL. Whereas PFOS and perfluorodecanesulfonate (PFDS) were the most abundant PFCs present in the nestlings, there was also evidence supporting decreases in the concentrations of PFOS, PFDS, and PFOA in the upper midwestern US. In addition, this research group argued that PFDS, which is a long-chained PFC that displays significant potential for bioaccumulation and subsequent toxicity, should be evaluated in detail in future wildlife toxicological studies (Route *et al.* 2014).

Furthermore, Vicente *et al.* (2015) determined the distribution of per- and polyfluoroalkyl substances (PFASs) in three-egg clutches of Audouin's gull (*Larus audouinii*) breeding in the colony in the Ebro Delta in the Southwestern province of Tarragona in Spain. Differential analysis of five PFASs in the yolk and albumen of 30 eggs revealed that PFOS was the main analyte detected, followed by perfluorononanoic acid (PFNA), perfluorohexane sulfonate (PFHS), and PFOA. Interestingly, this research group showed that the PFOS concentration decreased within the analyzed egg in accord with the sequence that the gulls' eggs were laid (Vicente *et al.* 2015).

The presence of PFASs in soft tissues (namely, liver, muscle, preen gland, and adipose tissues) and tail feathers of the terrestrial barn owl (*Tyto alba*) in Belgium, collected following accidental road kill (n = 15), was studied in depth by Jaspers *et al.* (2013). The authors reported that a major PFAS-manufacturing facility is located in the Antwerp area of Belgium. They also reported that previous studies had found very high levels of PFASs in the biota from that area (Jaspers *et al.* 2013). PFOA was found at high levels in the feathers (<14–670 ng/g wet weight) but not in the tissues of this species. The authors suggested that the detection of high levels of PFOA in barn owl feathers indicates that PFOA may be present on the surface of feathers because of external contamination from the air of point sources – a possibility that would permit the use of feathers for passive air sampling of high PFOA levels (Jaspers *et al.* 2013).

SUMMARY OF AMPHIBIAN TOXICOLOGY

There are limited studies available of the toxicological effects of PFOA on amphibians from the global literature, and these are largely restricted to studies conducted on *Xenopus* (South African clawed toads) embryos (Kim *et al.* 2013, Gorrochategui *et al.* 2016). One study compared the toxicity of different PFCs on vertebrate embryogenesis, using *Xenopus* embryos as a model system (Kim *et al.* 2013). A comprehensive analysis of teratogenic indices and the mechanisms underpinning developmental toxicity found that all PFCs tested were developmental toxicants and teratogens in *Xenopus*. Moreover, this research group suggested that the toxicological effects of PFCs increased as the fluorinated carbon chain increased (Kim *et al.* 2013). Although PFOA was shown to be a potent developmental toxicant, both perfluorodecanoic acid (PFDA) and perfluoroundecanoic acid (PFuDA) were found to be more potent developmental toxicants and teratogens in this animal model as compared to the other PFCs, due, in part, to these compounds containing a larger number of fluorinated carbons. This study also identified the heart and liver as primary developmental targets of PFOA, PFDA, and PFuDA (Kim *et al.* 2013).

A complementary study explored the effects of four PFAS compounds (namely PFBS, PFOA, PFOS, and PFNA) in *Xenopus laevis* A6 kidney epithelial cells by attenuated total reflection Fourier-transform infrared (ATR-FTIR) spectroscopy and chemometric analysis (Gorrochategui *et al.* 2016). This group noted a dose-dependent effect of PFASs on the growth and differentiation of A6 kidney epithelial cells, which was associated with the differential cytotoxicity of those compounds. The half-maximal effective concentration (EC_{50}) values of key PFAS compounds varied in order of decreasing concentration PFOS < PFNA < PFOA < PFBS, with higher concentrations of these PFASs provoking a much greater effect. Both PFOS and PFOA induced decreased A6 epithelial cell numbers, compared with the controls (Gorrochategui *et al.* 2016).

SUMMARY OF REPTILIAN TOXICOLOGY

Toxicological data for the effects of PFOA on reptiles is in very short supply. In the one study that explored the toxicological effects of PFCs in reptiles, Guerranti *et al.* (2013) determined the distribution of both PFOS and PFOA in blood specimens collected from loggerhead turtles, *Caretta caretta,* following a minimally invasive procedure. This research group obtained 49 blood specimens that were taken from groups of Italian marine turtles at the five designated rescue centers in Italy (i.e., n = 26 from Lesina; n = 13 from Linosa; n = 5 from Livorno; and n = 5 from Talamone for a total number of 49 collected blood samples) . In this study, PFOA was not detected, although PFOS was found at measurable levels in 15 of the 49 blood specimens at concentrations that ranged from 1.14 ng/g to 28.51 ng/g (wet weight). This research group did not find any marked differences between concentrations in specimens that were taken from different locations (Guerranti *et al.* 2013).

RECOMMENDED TOXICITY REFERENCE VALUES (TRVS)

TOXICITY REFERENCE VALUES FOR MAMMALS

TRVs for Ingestion Exposures for Mammalian Species

Dean and Jessup (1978) reported an oral lethal dose for 50 percent of the animals (LD_{50}) of 680 mg/kg and 430 mg/kg PFOA for male and female SD rats, respectively. Other groups have reported an oral LD_{50} of more than 500 mg/kg in male SD rats and between 250 and 500 mg/kg in female SD rats (Glaza 1997). In a similar study, Gabriel (1976a) reported an oral LD_{50} of less than 1,000 mg/kg for male and female Sherman-Wistar rats (Gabriel, 1976a). When considering these LD_{50} values in the context of the Hodge and Sterner toxicity scale, data support the concept that PFOA is moderately toxic after acute oral exposure (Gabriel 1976a). Liver toxicity and weight loss were the most commonly reported effects attributed to acute exposure to PFOA effects reported at concentrations as low as 1 mg/kg in mice (DeWitt *et al.* 2008, Eldasher *et al.* 2013, Wolf *et al.* 2008). Many of the studies involved a single dose, so that NOAEL values could not be established in most studies.

Sub-chronic PFOA oral gavage studies with rats and mice ranged in duration from six days of dosing during gestation up to 13 weeks of dosing (Table 2.3). Mortality in adults dosed with PFOA was rarely reported. Staples *et al.* (1984) reported that dosing female rats for ten days caused deaths in three of 25 individuals. Increased mortality in offspring from dosed adult females was reported when pregnant females were dosed at 30 mg/kg BW (Butenhoff *et al.* 2004). Effects on livers were reported at doses as low as 0.49 mg/kg-d in male mice (Son *et al.* 2008), and impaired mammary gland development was reported in mice at doses as low as 0.3 mg/kg-d (Dixon *et al.* 2012, Macon *et al.* 2011). One study reported liver effects (i.e., increases in absolute and relative liver weights and increased liver enzyme synthesis) at 0.64 mg/kg-d, and a NOAEL was established at 0.06 mg/kg-d in a 13-week study (Perkins *et al.* 2004).

Chronic low-dose PFOA exposures for two years in rats and mice had little impact on reproduction; however, gestational PFOA exposure induced delays in mammary gland development and/or lactational differentiation across three generations. In addition, chronic, low-dose PFOA exposure in drinking water was also sufficient to alter mammary gland morphological development in offspring of PFOA-exposed females, which was found to be visibly delayed from weaning onwards (White *et al.* 2011b), and induction of liver effects in males (Butenhoff *et al.* 2004) at 1 mg/kg-d. In each of these studies, a NOAEL could not be established. Proposed ingestion TRVs are presented in Table 2.6.

The above TRVs were selected for sensitive immunotoxicological end points for the class Mammalia (DeWitt *et al.* 2008). The second-order polynomial model was selected as the most representative model of the data. Benchmark dose (BMD)

TABLE 2.6

Selected Ingestion TRVs for PFOA for Mammalian Species

TRV	Dose (mg/kg-d)	Confidence
TRV-$_{LOW}$	1.75	Medium
TRV-$_{HIGH}$	3.06	Medium

Source: DeWitt et al. (2008).
TRVs were determined by the BMD/BMDL Benchmark Dose approach.

analysis of IgM responses (i.e., relative immune suppression) was derived from DeWitt *et al.* (2008), which gave a lower-bounded 95 percent confidence limit of 1.75 mg/kg-d on a BMD (at one SD about the mean) level of 3.06 mg/kg-d. Of note, the results of the DeWitt *et al.* (2008) study of PFOA dose-dependent IgM and IgG responses, as with similar data reported by Yang *et al.* (2000, 2001, 2002), all indicate that the immune system is a target of PFOA.

TRVs for PFOA Inhalation Exposures for the Class Mammalia

A single acute toxicity study of PFOA is reported for inhalation exposure. The study reports an LC_{50} in rats of 980 mg/m^3 (Kennedy *et al.* 1986). Mortality was observed at exposures as low as 380 mg/m^3 (the lowest concentration tested). Clinical signs of toxicity observed during the study are gasping, irregular breathing, and a red discharge around the eyes and nose (Kennedy *et al.* 1986).

Sub-chronic inhalation studies with rats ranged in duration from ten days to two weeks (Table 2.4). Mortality was reported at exposure concentrations as low as 25 mg/m^3. Effects on the liver were reported at 84 mg/m^3 (Kennedy *et al.* 1986). In addition, female rats were exposed to PFOA dust concentrations at 0.1, 1, 10, or 25 mg/m^3 *via* inhalation for 6 hours per day at GD6–15 (Staples *et al.* 1984). Three females (n = 6 per group) that were exposed to the 25 mg/m^3 dose died.

Clinical signs of toxicity in dams exposed to 10 and 25 mg/m^3 included wet abdomens, which began in the pelvic area, blood-stained tears and nasal discharge, and an unkempt appearance. In addition, food consumption was reduced in female rats that were exposed to 10 or 25 mg/m^3. No chronic inhalation studies have been published. Proposed inhalation TRVs are presented in Table 2.7.

Because no chronic studies were available and adverse effects were noted in sub-chronic bioassays conducted by Staples *et al.* (1984), with NOAELs of 1 mg/m^3 reported in two different studies (Kennedy *et al.* 1986; Staples *et al.* 1984), the sub-chronic results were used to develop the TRVs. The TRV Protocol (USACHPPM 2000) provides uncertainty factors of 10 for the NOAEL TRV and no uncertainty factors for the LOAEL TRV, when the TRVs are based on a sub-chronic NOAEL. Use of these uncertainty factors and the NOAEL of 1 mg/m^3 results in a NOAEL TRV of 0.1 mg/m^3 and a LOAEL TRV of 1 mg/m^3.

TABLE 2.7

Selected Inhalation TRVs for PFOA for Mammalian Species

TRV	Dose (mg/m³)	Confidence
TRV-$_{LOW}$	0.10	Medium
TRV-$_{HIGH}$	1.00	Medium

Source: Kennedy et al. (1986), Staples et al. (1984).
TRVs were determined by the NOAEL/LOAEL approach.

TOXICITY REFERENCE VALUES FOR BIRDS

Several studies have indicated that PFOA impacts health in birds by affecting the development of bird embryos and hatchlings (Smits and Nain 2013, Jiang *et al.* 2012, Strömqvist *et al.* 2012, O'Brien *et al.* 2009, Yanai *et al.* 2008, Pinkas *et al.* 2010). However, these studies reported toxicological end points, following largely *in ovo* delivery of PFOA into the air cell. In the context of the aforementioned studies, it is impossible to extrapolate for impacts of PFOA exposure to wildlife, and thus to establish a TRV, due to the absence of exposure data and concentrations from dosed adults. Therefore, insufficient data are available to determine a TRV for birds at this time.

TOXICITY REFERENCE VALUES FOR AMPHIBIANS

Limited studies of the toxicological effects of PFOA on amphibians were identified following our broad search of the available literature, with published studies being largely restricted to studies conducted on *Xenopus* embryos (Kim *et al.* 2013, Gorrochategui *et al.* 2016). One study evaluated the toxicity of PFCs toward vertebrate embryogenesis, using *Xenopus* embryos as a model system (Kim *et al.* 2013). In addition, a complementary study explored the effects of four PFAS compounds (PFBS, PFOA, PFOS, and PFNA) on *X. laevis* A6 kidney epithelial cells by attenuated total reflection Fourier-transform infrared (ATR-FTIR) spectroscopy and chemometric analysis (Gorrochategui *et al.* 2016). However, in the context of the aforementioned studies, it is impossible to extrapolate for impacts of PFOA exposure to wildlife and thus establish a TRV, due to the absence of exposure data and concentrations from dosed adult amphibian species. Therefore, insufficient data are available to determine a TRV for amphibians at this time.

TOXICITY REFERENCE VALUES FOR REPTILES

Toxicological data for the effects of PFOA on reptiles is severely limited. In a rare study that explored the toxicological effects of PFCs in reptiles, Guerranti *et al.* (2013) aimed to determine the distribution of both PFOS and PFOA in blood specimens that were derived from the loggerhead turtle (*Caretta caretta*) following a minimally invasive procedure. However, those data were of limited use to extrapolate to identify the impacts of PFOA exposure to reptiles and thus establish a TRV, due to the absence of exposure data and concentrations from dosed adult reptiles at this time.

IMPORTANT RESEARCH NEEDS

The lack of data on the toxicity of PFOA to wildlife species blocks the development of a TRV. Hence, more toxicological studies of the compound and its derivatives are recommended – particularly for poorly studied species, such as birds, reptiles, and amphibians. Most of the acute toxicity studies failed to establish NOAELs because they employed only a single dose. More studies with multiple doses are required to establish a suitable dose-response curve. In addition, most acute studies did not include sufficiently high doses to cause mortality, so the doses at which mortality would occur were not well defined. Studies of the effects of chronic PFOA exposure in mammals identify body weight loss as a key outcome (Butenhoff *et al.* 2004, 2012; Biegel *et al.* 2001), with body weight loss representing a highly relevant toxicological end point, especially under conditions where there was no decrease in food intake, and yet body weight decreased in neonates (Butenhoff *et al.* 2004).

Since many of these studies were designed to involve single-dose exposures, NOAEL and LOAEL TRVs could not be established, and thus the lower limits of exposure that will lead to chronic adverse effects remain poorly understood. Clearly, there is a significant research need for additional dose-response studies in mammals and particularly across a wider range of species within the Class. Only one study (Tucker *et al.* 2015) compared response of different genotypes to PFOA, estimating a NOAEL for PFOA from studies conducted on CD-1 (NOAEL = 0.01 mg/kg-d) as compared with C57Bl/6 (NOAEL = 0.1 mg/kg-d) mice, using an experimental approach that was aimed at exploring the toxicological effects of PFOA on mammary gland development.

Data from inhalation and dermal toxicity testing is limited for laboratory mammals and completely lacking for all other groups. Limited data are available for amphibians and reptiles, whereas the publications on the few avian species which have been studied are restricted to egg-injection studies. Studies that focus on both acute and chronic toxicity studies on wild mammals as well as on non-mammalian wildlife, such as birds, reptiles, and amphibians, would be particularly warranted. Further studies are needed to identify the novel pathways regulating the endocrine-disrupting effects of PFOA and the doses at which they are active and relevant (White *et al.* 2011a). Similarly, insights into the immunotoxicity of PFOA and its potential to suppress host immunity and increase susceptibility to infectious diseases require additional studies to determine the mechanisms of action and identify the species at risk (DeWitt *et al.* 2008, 2009, Qazi *et al.* 2012, Fair *et al.* 2013, Smits and Nain 2013).

3 Perfluorooctane Sulfonate (PFOS)

Marc A. Williams

CONTENTS

PERFLUOROOCTANE SULFONATE AND ITS USES

A variety of manufactured per- and polyfluoroalkyl substances PFAS compounds, including perfluorooctane sulfonate (PFOS), have found broad utility in many consumer goods, and are widely recognized as emerging pollutants with global impact with regard to their diverse toxicological effects (Vierke *et al.* 2012, NTN 2015, USEPA 2014b, 2016a). Since the early 1950s, the unique chemical properties of PFASs have resulted in their extensive use in surface coatings and protectant formulations (Buck *et al.* 2011). Specific examples include consumer goods (e.g., nonstick kitchenware, waterproof clothing, and cosmetics), and industrial products (e.g., firefighting foams, industrial surfactants, emulsifiers, wetting agents, industrial additives, and coatings; 3M 2000a, Schultz *et al.* 2003, ATSDR 2015). Extensive use and potential for exposure of humans and wildlife to materials and products containing PFASs has a long history. Furthermore, although PFAS emissions have dramatically reduced recently, the bioaccumulation and environmental persistence of perfluorinated compounds (PFCs) results in detectable levels in both wildlife and the human population (NTP 2016).

TOXICOLOGICAL EFFECTS OF PFOS ON WILDLIFE

This section represents a systematic and detailed assessment of the available scientific literature with the aim of determining the toxicological characteristics of perfluorooctane sulfonate (PFOS) that might have importance for the health of wildlife (i.e., wild mammals, birds, reptiles, and amphibians) potentially exposed to this chemical. This chapter provides a summary of the toxicological effects of PFOS on wildlife, following exposure, and evaluates PFOS toxicity data that is used to derive toxicity reference values (TRVs).

The TRVs are to be used as screening-level benchmarks for wildlife at or in close proximity to contaminated sites. The protocols for performing this assessment are, in part, described in Technical Guide No. 254 – the *Standard Practice for Wildlife Toxicity Reference Values* (USACHPPM 2000, Johnson and McAtee 2015, Deck and Johnson 2015).

ENVIRONMENTAL FATE AND TRANSPORT

All PFCs found in the environment are anthropogenic substances that do not occur naturally and are incredibly stable chemicals, although there is some evidence that partially fluorinated hydrocarbons can undergo chemical breakdown at functionally-bonded groups. PFASs are characterized by a fully (per-) or partly (poly-) fluorinated carbon chain that is linked to different functional groups. Two groups of perfluorinated compounds, known as perfluoroalkyl sulfonates (PFSAs) and perfluorocarboxylic acids (PFCAs), have attracted considerable attention due to their demonstrated toxicity, as detailed in the relevant sections that follow.

Prior uses of PFOS included industrial and domestic applications, like stain repellents, and water-repellent fabric, that were manufactured by the 3M Company.

Industrial manufacture of many C_8-PFC homologs has been restricted in many countries around the world. One example includes the notable action by the 3M Company – the primary manufacturer of perfluooroctyl sulfonyl fluoride (POSF) – which, in 2000, voluntarily phased-out production of POSF and has since ceased all production of it in the US (3M 2000a, 2000b, ATSDR 2009, 2015, USEPA 2006, 2014b, Olsen *et al.* 2003).

PFOS and PFOA, both of which have eight carbon chain lengths, remain the predominant PFASs, that are ubiquitously found in water columns, biota, and sediments (Ahrens *et al.* 2009, Butt *et al.* 2010, Kwadijk *et al.* 2010). This is due to widespread use and emissions *via* industrial, military, and firefighting operations. It is also due to their detection in and around PFAS manufacturing plants, their migration out of consumer products into the air, their presence in household dust, and in food and soil, and from ground and surface water, from which they make their way into drinking water (Darwin 2011, Fire Fighting Foam Coalition 2014, Shoeib *et al.* 2006, Tittlemier *et al.* 2007, Trier *et al.* 2011, Sepulvado *et al.* 2001, Strynar *et al.* 2012, Eschauzier *et al.* 2012, Rahman *et al.* 2014).

PFOS ($C_8F_{17}SO_3^-$), like PFOA ($C_7F_{15}COO^-$), has known toxicology, and, like many PFASs, its high-energy perfluorinated chain means that it degrades very slowly under typical biotic and abiotic environmental conditions and persists in the environment (Conder *et al.* 2010). Indeed, PFOS is resistant to thermal, chemical, biological, and microbial degradation, and metabolism by vertebrates, and is quite resistant to hydrolysis, direct photolysis, and atmospheric photo-oxidation (ATSDR 2009). Thus, PFOS persists in the environment and is capable of long-range transport (Lau *et al.* 2007, EFSA 2008, ATSDR 2009, 2015, USEPA 2014a–c, 2016a,b). Consequently, PFOS, like PFOA (Chapter 2), poses a high risk of disseminated global environmental contamination, and is commonly found in wildlife and human organs, tissues, and bodily fluids (Buck *et al.* 2010, 2011, Ahrens 2011, Sturm and Ahrens 2010, Houde *et al.* 2006a, Butenhoff *et al.* 2006, Calafat *et al.* 2007, Giesy and Kannan 2001, Kannan *et al.* 2004).

The unique chemical properties of organofluorine molecules arise from the distinct properties of fluorine, the most electronegative element in the periodic table, which attracts electrons in a chemical bond towards itself, conferring high polarity and strength to the carbon–fluorine bond (i.e., 110 kcal mol^{-1}). Due to the surface-active properties of PFOS, it forms three layers in octanol/water, making an n-octanol/water (K_{ow}) partition co-efficient impossible to determine experimentally. When attached to a perfluorinated chain, a charged moiety (e.g., sulfonic acid) imparts hydrophilicity to the structure. Thus, functionalized fluorochemicals, like PFOS, have surfactant properties, and selectively adsorb at interfaces due to the display of both hydrophobic and hydrophilic moieties.

In terms of abiotic and biotic degradation, PFOS shows no evidence of hydrolysis, with an estimated half-life of more than 41 years at 25°C and no evidence of photolysis, with an estimated half-life of more than 3.7 years, with none of the studies surveyed showing evidence of biodegradation of PFOS in the aquatic environment under aerobic or anaerobic conditions (UK Environment Agency 2004). Others have reported that some polyfluorinated chemicals might break down to form

perfluorinated chemicals in the environment (D'Eon and Mabury 2007). For example, PFOS precursors might degrade and liberate PFOA (USEPA 2014b). In addition, both PFOA and PFOS have great potential for biomagnification in food webs
(Martin *et al.* 2004a, Kelly *et al.* 2009, Loi *et al.* 2011). However, it is uncertain what
constitutes the major route of PFOS transport to remote locations. This uncertainty
is complicated by the fact that PFOS almost completely ionizes and is less volatile
in this form.

PFOS is commonly used as a simple salt (e.g., as the potassium, sodium, or ammonium salt) or incorporated into larger polymers (EFSA 2008, USEPA 2009). Low
log_{10} acid dissociation constants (pK_a), ranging from −3 to +4, suggest that PFOS
and PFOA are strong acids and it is expected that both will exist predominantly in
the anionic form in the environment (Kissa 2001, ATSDR 2009, Conder *et al.* 2010).
Volatilization of perfluoroalkyl anions, such as perfluorooctanoate (PFO), from water
surfaces is expected to be negligible since ions do not volatilize (Prevedouros *et al.*
2006). However, due to the surfactant nature of the perfluoroalkyl compounds, some
of the perfluoroalkyl anions that are released to water may form micelles and exist in
the associated form, despite their low pK_a values (USEPA 2005a; Prevedouros *et al.*
2006). Perfluoroalkyl compounds that associate on water and soil surfaces may also
volatilize into the atmosphere (USEPA 2005a, Kim and Kannan 2007).

Additionally, when released directly to the atmosphere, PFCs are expected to
adsorb to particles and settle onto soil through wet or dry deposition, representing
the principal removal mechanisms for these compounds in particulate form from the
atmosphere (Barton *et al.* 2006, ATSDR 2009). In addition, residence times with
respect to these processes would extend from days to weeks (Barton *et al.* 2006,
Hurley *et al.* 2004, Kim and Kannan 2007). PFOA and PFOS in anionic form are
water soluble and can migrate readily from soil to groundwater (Conder *et al.* 2010,
Post *et al.* 2012). Due to their chemical stability and low volatility in ionic form,
PFCs are persistent in water and soil (ATSDR 2009).

It is now generally appreciated that PFOS is present in many disparate environmental media, including the aqueous phase, examples of which include wastewater
(Zareitalabad *et al.* 2013), surface water (Zareitalabad *et al.* 2013, Moody *et al.* 2001,
2002, Hansen *et al.* 2002, Lindim *et al.* 2016, Yu *et al.* 2013), groundwater (Chen
et al. 2016, Liu *et al.* 2016), and tap water (Jin *et al.* 2009). PFOS was also detected
in the solid phase, where many studies have explored the sorption of PFOS to sediments (Perra *et al.* 2013, Chen *et al.* 2012, Shu Chen *et al.* 2016, Huiting Chen *et
al.* 2016, Higgins and Luthy 2006, Kwadijk *et al.* 2013, Pan *et al.* 2009), soil (Shu
Chen *et al.* 2016, Huiting Chen *et al.* 2016, Strynar *et al.* 2012), minerals (Zhao *et al.*
2014), and sludge (Milinovic *et al.* 2016, Wang *et al.* 2015b), as well as in the ambient air (Goosey and Harrad 2012). Additionally, perfluoroalkyl carboxylic acids and
sulfonic acids have been detected widely in both environmental media and biota of
the Arctic regions and in other remote locations.

At many military bases and installations in the US, including Tyndall Air Force
Base (AFB), near Panama City in Florida, and Wurtsmith AFB in Michigan, fluoro-
chemical-contaminated groundwater plumes associated with past fire-fighting train-
ing sites were found (Levine *et al.* 1997, Moody *et al.* 2003). Perfluorocarboxylate

concentrations of 125–7,090 µg/L were found in the groundwater at Tyndall (Moody *et al.* 2003). It was previously shown that surface water samples collected near a spill site of aqueous film-forming foam (AFFF) concentrate contained levels of per-fluorohexane-sulfonate (PFHS), PFOS, and PFOA at concentrations of 0.011–2,270 µg/L (Moody *et al.* 2003). AFFFs are very effective at extinguishing liquid fuel fires and are widely and routinely used by both civilian and US military fire-fighters (Darwin *et al.* 1995, Chan and Chian 1986). With accumulating and emergent data, it is now appreciated that PFOS in contaminated soils has the potential to leach into groundwater and contaminate tap/drinking water reservoirs and surface water, with an equal likelihood of it being taken up by environmental receptors that include plants and soil organisms (Hellsing *et al.* 2016).

Sorption/desorption behaviors (i.e., partitioning between the aqueous and solid phases) of PFOS and the associated mechanisms involved in these processes were studied by Wei *et al.* (2007). In the current study, the sorption and desorption of PFOS onto six soil types, displaying different physico-chemical properties, were investigated. Kinetic and equilibrium studies of PFOS sorption onto the soil types were conducted by batch analysis. It was found that PFOS sorption equilibrated within 48 hours. Moreover, fitting the experimental data to models of pseudo-second-order rate kinetics showed that PFOS sorbed onto soil samples by chemisorption (Wei *et al.* 2007). Moreover, the intraparticle diffusion model data indicated that both film diffusion and intraparticle diffusion were rate-limiting steps for five of the six soil samples, while intraparticle diffusion was the sole limiting step in PFOS sorption to the sixth soil type (Wei *et al.* 2007). It is now recognized that one of the most important processes that affects the fate and transport of PFOS among many different environmental media is its partitioning between aqueous and solid phases (i.e., the kinetics of sorption/desorption).

BIOACCUMULATION AND ELIMINATION

The wide distribution of PFCs in high trophic levels increases the potential for bioaccumulation and bioconcentration. The bioaccumulation potential of perfluoroalkyls is reported to increase as the chain length increases by four to eight carbons, with subsequent decline with further increases in chain length (de Vos *et al.* 2008, Furdui *et al.* 2007, Martin *et al.* 2004b, ATSDR 2015). PFOS and other perfluoroalkyl compounds are absorbed by and eliminated from animal species, with differential measured half-lives. In living organisms, perfluoroalkyls bind to the protein albumin in the blood, liver, and eggs but do not accumulate in fat tissue (de Vos *et al.* 2008, Kissa 2001).

The half-lives determined for primates, rat, and mouse were 121, 48, and 37 days, respectively, whereas, in humans, it was reported to be 5.4 years. The mean PFOS concentration reported in various studies of the US general population was estimated to be 14.7 to 55.8 ng/L. In addition, it was previously reported that the highest bioaccumulation in rats of PFOS was found in the liver (648 µg/g) following exposure of rats to 20 mg/kg-d for 28 days (Cui *et al.* 2009). Perfluoroalkyl compounds and PFOS were detected at concentrations of 121 to 235 ng/ml in serum samples collected from domestic cats in veterinary clinics (Bost *et al.* 2016).

Although perfluoroalkyl compounds, like PFOS, exist in the environment as complex and structurally mixed linear and branched forms, there is a paucity of data with regard to their differential and structure-dependent bioavailability, and the mechanism whereby PFOS and other perfluoroalkyl compounds are taken up by animals remains unclear (de Vos *et al.* 2008, Kissa 2001). Nevertheless, concentrations of perfluoroalkyl compounds have been measured in invertebrates, fish, amphibians, reptiles, birds, the eggs of birds, and many globally distributed species of mammals, and, of course, in humans (Dai *et al.* 2006, Giesy and Kannan 2001a, Houde *et al.* 2005, 2006a, 2006b, Keller *et al.* 2005, Kannan *et al.* 2001a, 2001b, 2002a, 2002b, 2002c, 2002d, 2005, 2006; Sinclair *et al.* 2006, So *et al.* 2006, Wang *et al.* 2008). The highest concentrations of individual perfluoroalkyls in animals have been measured in apex predators, including the polar bear (de Vos *et al.* 2008, Houde *et al.* 2006b, Kannan *et al.* 2005, Kelly *et al.* 2009). This informs us that some perfluoroalkyls can bioaccumulate in food webs (de Vos *et al.* 2008, Houde *et al.* 2006a, Kannan *et al.* 2005, Smithwick *et al.* 2005a, 2005b, 2006, Kelly *et al.* 2009). However, perfluoroalkyl sulfonates with carbon chain lengths less than eight do not bioaccumulate as well as PFOS.

Due to the persistence and capacity of PFOS for long-term accumulation, higher trophic level wildlife like fish, piscivorous birds, and Arctic biota are at risk of continuous exposure to PFOS and PFOA (USEPA 2006, UNEP 2006). Long-range sources of PFCs include the atmospheric transport of precursor compounds, including perfluoroalkyl sulfonamides, and direct long-range transport of PFCs *via* ocean currents or in the form of marine aerosols (ATSDR 2009, Post *et al.* 2012). PFOS also has a greater tendency than PFOA to bind to organic matter and to bioaccumulate, due to its longer perfluoroalkyl chain length (Conder *et al.* 2010).

Furthermore, PFOS has been detected in fish-eating birds such as the bald eagle and the albatross (Kannan *et al.* 2001). Falk *et al.* (2012) analyzed pooled livers of roe deer (*Capreolus capreolus*) that were collected from 1989 to 2010 and showed that the highest quantified levels of PFOS had a median level of 6.3 µg/kg. By contrast, PFOA, perfluorononanoic (PFNA), and perfluorodecanoic acid (PFDA) were detected at concentrations of 0.5, 1.2, and 0.3 µg/kg, respectively. In addition, both PFOS and PFOA were detected in wild boar (*Sus scrofa*) tissues at concentrations of 178 µg/kg and 45 µg/kg, respectively, in the liver, and at 28 µg/kg and 7.4 µg/kg, respectively in the muscle tissue (Stahl *et al.* 2012).

In terms of elimination of PFOS, studies have revealed significant inter-species variation in the elimination rates of PFOS from animals (Hundley *et al.* 2006). Elimination of PFOS from male rats (half-life ($t_{1/2}$) of approximately 180 hours) was approximately 20 times faster than was seen in crab-eating macaques ($t_{1/2}$ approximately 150 days), and approximately 240 times faster than was seen in humans ($t_{1/2}$ approximately 1,800 days). There are important implications that should be considered when taking the elimination rates into account. For example, similar exposure doses of PFOS (i.e., quantified in mg/kg-d) to rats, monkeys (or, indeed, humans) will elicit markedly different steady-state internal doses (i.e., body burden and serum concentrations) of PFOS, depending on the species of interest. Also, exposure durations that are needed to achieve the steady-state situation would be expected to be significantly more prolonged in humans than in monkeys or rats.

Additionally, the time taken to achieve 95 percent steady-state would be expected to vary markedly between species and between genders within a species. In the example of perfluorobutyric acid, interspecies variability of the elimination rate exceeded that expected for allometric scaling of body weight (Chang *et al.* 2008). In addition, the monkey: rodent ratio for an elimination half-life was approximately 14 for female mice and 20–40 for female rats; based on body weight alone, one would predict ratios of approximately 4.5 and 2.5, respectively. In addition to faster elimination rates of PFOA and PFOS from female rats than male rats, elimination rate and renal clearance of PFOA from female rats appeared to be dose dependent (Kemper 2003). Dose- and gender-specific differences in clearance rates vary significantly between species. Furthermore, extrapolation of external dose dependency or serum concentration dependency of the elimination rates of PFOA or PFOS observed in rats to other species would carry a high degree of uncertainty (ATSDR 2009).

Toxicokinetic studies by Chang *et al.* (2012), in which a single oral dose of PFOS-^{14}C in solution at 4.2 mg/kg was administered to male rats (n = 3), found that 48 hours after dosing, 3.32% of the total dose was located in the gastrointestinal tract and 3.24% in the feces. Observations suggested that the bulk of the administered dose was absorbed with a small fraction of the unabsorbed material excreted in fecal matter.

In complementary work, Martin *et al.* (2007) administered 10 mg PFOS/kg to adult male Sprague Dawley (SD) rats (n = 5) for 1, 3, or 5 days by oral gavage and subsequently determined the liver and serum levels. Blood was collected *via* cardiac puncture and the concentration of PFOS was determined by high-performance liquid chromatography-electrospray tandem mass spectrometry. The mean (± standard error) liver PFOS concentrations after 1, 3, or 5 daily doses were 83 ± 5, 229 ± 10, and 401 ± 21 µg/g, respectively. The mean serum concentration after 1 or 3 daily doses was 23 ± 2.8 and 87.7 ± 4.1 µg/mL, respectively.

In an adult male C57BL6 mouse model, PFOS-^{35}S was administered in the feed at low or high doses for 1, 3, or 5 days (n = 3 mice/group), wherein the dose equivalent for low-dose exposures was 0.031 mg/kg-d, and, for the high-dose exposures, the dose equivalent was 23 mg/kg-d (Bogdanska *et al.* 2011). In high-dose exposures after 5 days, mice presented with hypertrophy of the liver, and atrophy of the fat pads and epididymal fat when compared to their low-dose exposure counterparts at 5 days. Additionally, by measuring the hemoglobin concentration in all of the samples, the amount of radioactivity recovered, due to blood in the tissues, was determined. By correcting for PFOS in the peripheral blood, the actual tissue levels were determined (Bogdanska *et al.* 2011). At both the low and high doses, and at the 1-, 3-, or 5-day time points, the liver displayed the highest concentrations of PFOS. At the low dose, the liver PFOS level, relative to the blood concentration, increased with time, whereas, at the high dose, the ratio plateaued after three days. At the high-dose exposures, the liver contained 40–50% of the recovered PFOS. It was thus suggested that this could reflect high levels of PFOS binding to tissue proteins. In high-dose mice, the next highest level of PFOS was found in the lungs. With the exception of the liver, the tissue-to-blood ratio for the lung exceeded that of all other tissues studied. The lowest PFOS levels were found in brain and fat deposits. Furthermore, studies by whole-body autoradiography of a mouse 48 h, after a single oral dose of PFOS-^{35}S (12.5 mg/kg), indicated that the bone marrow, not the calcified bone, contained the majority of the deposited PFOS (Bogdanska *et al.* 2011). Finally, levels of PFOS in the kidney approximated to those values seen in the peripheral blood at both dosages and all time points.

MAMMALIAN ORAL TOXICITY

MAMMALIAN ORAL TOXICITY – ACUTE AND SUB-ACUTE

In toxicity studies, several lines of evidence indicate health effects following exposure to PFOS. Table 3.1 provides a summary of the acute and sub-acute oral toxicology of PFOS.

When dosed by oral gavage, sub-acute toxicity studies in Sprague Dawley (SD) rats (n = 5 per gender per dose) at single doses of 0, 100, 215, 464, or 1,000 mg/kg of PFOS, suspended in 20% acetone/80% corn oil, revealed 100% mortality at the highest doses of 464 and 1,000 mg/kg, with 30 percent of the rats in the 215 mg/kg group failing to survive across a 14-day sub-acute study in which rats were observed for abnormal effects for 4 hours after exposure and then daily, up to 14 days. Clinical signs included hypoactivity, decreased limb tone, and ataxia. Necropsy indicated stomach distention, lung congestion, and irritation of the glandular stomach mucosa. Based on the dose-response findings, the acute oral LD_{50} was 233 mg/kg in males, and 271 mg/kg in females. The combined-gender LD_{50} was 251 mg/kg in rats (Dean *et al.* 1978).

In another 14-day sub-acute study, male Wistar rats and male ICR mice (n=2–3 per dosed group) were administered a single oral dose of PFOS at concentrations of 0, 125, 250, or 500 mg/kg, and monitored for any neurological outcomes (Sato *et al.* 2009). This study revealed decreased or delayed body weight gains in rats of the 250 mg/kg dosed group, and mortality was seen in rats of the 250 mg/kg (1/3) and 500 mg/kg (2/2) dosed groups (Sato *et al.* 2009). In these same rats, the highest concentration of PFOS was detected in the liver and the lowest was in the brain. Rats administered 250 mg/kg body weight (BW) did not show any differences in the levels of catecholamines (i.e., norepinephrine, dopamine, or serotonin) or amino acids (i.e., glutamic acid, glycine, and gamma-aminobutyric acid [GABA]) when compared with the controls at 24 and 48 hours post-exposure (Sato *et al.* 2009). Oral toxicity following PFOS exposure at single oral doses of PFOS at 0, 125, 250, or 500 mg/kg was also explored in male ICR mice (2–3 mice per group). This study showed decreased body weight gains during the 14-day observation period for the 250 mg/kg dosed group, and one mouse in each PFOS exposed group failed to survive treatment as compared the control (0 mg/kg (BW) PFOS) group. No neurological signs were found in treated mice. No histopathological changes were seen in glial or neuronal cells of the cerebrum and cerebellum in rats killed 24 hours post-exposure (Sato *et al.* 2009). Another group explored the effects of orally exposed male mice, to which PFOS was administered orally at doses of 0, 5, 20, or 40 mg/kg for 7 days (Zheng *et al.* 2009). Observations included dose-dependent decreases in food intake and body

TABLE 3.1
Summary of Acute Oral Toxicity for PFOS in Mammalian Species

Test Organism	LD_{50} (mg/kg)	NOAEL (mg/kg-d)	LOAEL (mg/kg-d)	Test Results — Effects Observed at the LOAEL	Reference
Monkey	NA	NA	10	Death at all dose levels	Goldenthal et al. 1978
Mouse	NA	NA (natural killer (NK) cells)	5.0 (SRBC) 20 (NK Cells)	Immunotoxicity. Increased liver weights and suppression of the plaque-forming cell response (sheep red blood cells, SRBC); Impaired T cell function. LOAEL also based on decreased NK cell activity	Zheng et al. 2009
Mouse	NA	NA	5.0	Based on weight loss on a high-fat diet	Wang et al. 2014
Mouse	NA	1.0	5.0	Based on increased liver weight, changes in oxidation biochemical parameters	Wan et al. 2012
Mouse	NA	0.3	3.0	Based on increased liver organ weights in dams and male offspring, and increased fasting serum insulin levels in males	Wang et al. 2014
Mouse	NA	NA	NA	Developmental neurotoxicity study. Mice at both concentrations tested (0.75 and 11.3 mg/kg) showed decreased physical activity and increased neuroprotein	Johansson et al. 2008, 2009
Mouse	NA	1.0	5.0	Authors calculated a $BLDL_5$ for survival at 6 days of 3.88 mg/kg-d. Based on increased liver weights, mouse pup mortality, and survival	Lau et al. 2003
Mouse	NA	1.0	5.0	Maternal LOAEL based on increased liver weight	Thibodeaux et al. 2003

(Continued)

TABLE 3.1 (CONTINUED)
Summary of Acute Oral Toxicity for PFOS in Mammalian Species

Test Organism	LD$_{50}$ (mg/kg)	NOAEL (mg/kg-d)	LOAEL (mg/kg-d)	Effects Observed at the LOAEL	Reference
				Test Results	
Mouse	NA	1.0	5.0	Developmental LOAEL based on increased liver weight and delayed eye opening. A BMDL5 of 3.88 mg/kg corresponded to the maternal dose regimen for survival of mouse pups at post-natal date (PND) 6	Lau et al. 2003
Mouse	NA	Males 0.1 Females 1.0	Males 1.0 Females 5.0	Developmental immuno-toxicity. Male NOAEL based on decreased NK cell activity at PND 8. Female NOAEL also based on decreased NK cell activity at PND 8	Keil et al. 2008
Mouse	<500	NA	NA	Increased mortality and decreased body weight	Sato et al. 2009
Mouse	NA	Maternal 6.0 Developmental NA	Maternal NA Developmental 6.0	Developmental toxicology. Developmental LOAEL based on decreased body weight	Fuentes et al. 2007
Mouse	NA	Maternal 1.0 Developmental 1.0	Maternal 10.0 Developmental 10.0	Developmental toxicology. Maternal LOAEL based on increased liver, organ weights. Developmental LOAEL based on observed fetal abnormalities and decreased rates of survival	Yahia et al. 2008

(Continued)

TABLE 3.1 (CONTINUED)

Summary of Acute Oral Toxicity for PFOS in Mammalian Species

Test Organism	LD$_{50}$ (mg/kg)	NOAEL (mg/kg-d)	LOAEL (mg/kg-d)	Effects Observed at the LOAEL	Reference
			Test Results		
Rat	Combined 251 Males 233 Females 271	NA	NA	Hypo-activity, decreased limb tone, ataxia. Gross necropsy revealed stomach distension and lung congestion	Dean et al. 1978
Rat	NA	NA	12.5	Decreases in body weight	Yang et al. 2009
Rat	NA	NA	NA	Total thyroxine (TT4) decreased at 2, 6, and 24 hours. Triiodothyronine (T3) and reverse triiodothyronine (rT3) – decreased at 24 hours. Free thyroxine increased at 2 and 6 hours, and normal at 24 hours	Chang et al. 2008
Rat	NA	NA	1.72	Liver effects - increases in liver weight, increases in serum alanine aminotransferase (ALT) and aspartate aminotransferase (AST) activities, and decreases in serum cholesterol levels after acute exposure	Elcombe et al. 2012b
Rat	500	NA	NA	100 percent mortality	Sato et al. 2009
Rat	NA	Maternal 1.0 Developmental 0.3	Maternal NA Developmental 1.0	Developmental LOAEL based on increased motor activity	Butenhoff et al. 2009

(Continued)

TABLE 3.1 (CONTINUED)
Summary of Acute Oral Toxicity for PFOS in Mammalian Species

Test Results

Test Organism	LD$_{50}$ (mg/kg)	NOAEL (mg/kg-d)	LOAEL (mg/kg-d)	Effects Observed at the LOAEL	Reference
Rat	NA	Maternal 1.0	Maternal 2.0	Maternal LOAEL based on decreased whole body weights	Thibodeaux et al. 2003
Rat	NA	Developmental 1.0	Developmental 2.0	Developmental LOAEL based on decreased survival BMDL$_5$ corresponding to the maternal dose for rat pup survival at PND 8 was 0.58 mg/kg	Lau et al. 2003
Rat	NA	Offspring NA	Offspring 0.1	LOAEL based on changes in the cortex and hippocampus (i.e., activation markers and pro-inflammatory transcription factors of astrocytes)	Zeng et al. 2011
Rat	NA	Offspring at PND 21 0.1	Offspring at PND 21 0.6	Based on increased programmed cell death in heart cells	Zeng et al. 2014
Rat	NA	Offspring 0.1	Offspring 2.0	Based on histopathological changes in the lungs, decreased body weight, and increased mortality	Chen et al. 2012
Rat	NA	NA	0.5	Based on decreases in offspring body weight, impaired glucose tolerance	Lv et al. 2013
Rat	NA	NA	5.0	Based on decreased offspring body weight	Zhao et al. 2014
Rat	NA	0.8	2.5	Developmental toxicology. LOAEL based on increased water maze escape distance and escape latency	Wang et al. 2015
Rabbit	NA	NA	NA	Dermal toxicology. Dose of 0.50 g/eye – no gender-specific data. No irritation seen Ocular toxicology. Dose of 0.50 g – no data on gender-specific differences. Exact score not provided, with the exception of the maximal score at 1 and 24 hours post-treatment	Biesemeier and Harris 1974

(Continued)

TABLE 3.1 (CONTINUED)
Summary of Acute Oral Toxicity for PFOS in Mammalian Species

Test Organism	LD$_{50}$ (mg/kg)	NOAEL (mg/kg-d)	LOAEL (mg/kg-d)	Test Results — Effects Observed at the LOAEL	Reference
Mouse	NA	Male 0.00017, Female 0.0033, Male 0.0033, Female 0.166	Male 0.0017, Female 0.017, Male 0.017, Female NA	Immunotoxicity study Based on decreased SRBC plaque-forming cell response Based on decreased SRBC plaque-forming cell response Based on decreased NK cell activity in females and increased NK cell activity in males	Peden-Adams et al. 2008
Mouse	NA	1.0	5.0	Based on increased liver organ weights and appearance of hepatic steatosis	Wan et al. 2012
Rat	NA	NA	NA	Based on decreased body weight and observations of lung congestion	Cui et al. 2009
Rat	NA	NA	NA	Based on decreased levels of luteinizing hormone (LH) and testosterone, and increased levels of follicle-stimulating hormone (FSH)	Lopez-Doval et al. 2014
Rat	NA	Males 0.37, Females 0.47	Males 1.51, Females 1.77	Based on decreased body weight in males and females, and decreased food consumption in females	Seacat et al. 2003
Rat	NA	Males 0.14, Females NA	Males 1.33, Females 0.15	Based on increased relative liver, organ weights	Curran et al. 2008

Key: NA = not applicable; PND = post-natal day; SRBC = sheep red blood cells weight loss, and dose-dependent increases in liver weight, that was accompanied by enhanced serological levels of corticosterone. In addition, dose-dependent decreases in splenic and thymic cellularity were seen, which were consistently associated with significant and dose-dependent decreases in absolute numbers of immunophenotypically classified thymic T cell sub-populations and decreases in the immunological and functional competence of T and B cells.

Others have exposed male SD rats to orally administered doses of potassium (K+) PFOS in the feed at 20 or 100 ppm for 1, 7, or 28 days (Elcombe et al. 2012a). Exposed rats presented with decreased mean body weight (100 ppm only, at 28 days), and increased mean relative liver weight by day 7 for the 100 ppm dose only, and by day 28 for both the 100 and the 20 ppm exposed groups (Elcombe et al. 2012a). In addition, in K+PFOS-exposed mice, mean plasma levels of cholesterol were decreased after 7 days (100 pm dose only) and after 28 days (both 20 and 100 ppm doses). Moreover, the same research group showed that, in SD rats orally exposed to K+PFOS in the diet, the plasma activities of the enzymes alanine aminotransferase (ALT) and aspartate aminotransferase (AST), were less than the corresponding control values at all of the time points measured, though these activity differences were generally not statistically significant (Elcombe et al. 2012b).

Curran et al. (2008) conducted two 28-day dietary exposure studies (n = 15 per group by gender) in SD rats by administering 0, 2, 20, 50, or 100 mg PFOS/kg diet, which was equivalent to 0, 0.14, 1.33, 3.21, and 6.34 mg/kg body weight (BW)/d respectively in males, and 0, 0.15, 1.43, 3.73, and 7.58 mg/kg BW/d respectively for females. In the first study, each rat was assessed for clinical and hematological parameters, and for changes in histopathology and gene expression patterns following dietary exposure to PFOS. In the second study, the intention was to explore the effects of dietary PFOS exposure on blood pressure, erythrocyte deformability, and liver fatty acid composition (Curran et al. 2008). The authors reported no meaningful treatment-related differences in the context of hematological or urinalysis end points. There were significant reductions in body weights in both male and female rats at days 21 and 28 following exposure to the 50 or 100 mg PFOS/kg diet, whereas, in male rats, food consumption was markedly reduced from week 2 through to week 4 in the 50 and 100 mg PFOS/kg diet groups.

In female rats, food consumption in the 100 mg/kg dose group was reduced, compared with controls in weeks 1, 3, and 4. In addition, at week 4 only, female rats in the 50 mg/kg dose group consumed less food than did the controls (Curran et al. 2008). In terms of histopathological changes, evidence of increased cytoplasmic homogeneity and hepatocyte hypertrophy was seen in both male and female rats at PFOS doses ≥ 50 mg/kg diet. With respect to clinical biochemical parameters, male and female rats responded similarly, with serum bilirubin concentration and alanine aminotransferase (ALT) activity being elevated, while cholesterol and triglyceride concentrations were reduced (Curran et al. 2008).

Finally, at high PFOS doses, the serum levels of conjugated and total bilirubin were increased. Serum cholesterol concentration was decreased for males and females on ≥ 50 mg PFOS/kg diet. Serum levels of the thyroxine (T_4) and triiodothyronine (T_3) tyrosine-based hormones that play critical roles in metabolism were also decreased in males and females, with T4 levels being statistically significantly lower at $\geq$ 20 mg PFOS/kg diet than the control levels (Curran *et al.* 2008). It was determined that the LOAEL was the 20 mg/kg dietary level (i.e., in males: 1.33 mg PFOS/kg-d; in females: 1.43 mg PFOS/kg/-d) to achieve a significant increase in absolute (females) or relative (males and females) liver weights and significant decreases in serum T4 concentration (males and females). The NOAEL was the 2 mg/kg diet level (i.e., 0.14–0.15 mg PFOS/kg-d).

Other groups showed that PFOS primarily accounted for dampened mitochondrial β-oxidation activity as monitored on day 14 (Wan *et al.* 2012). Whereas total and peroxisomal β-oxidation were slightly, but increased ($P < 0.01$) at a dose of 10 mg/kg-d, mitochondrial β-oxidation was markedly decreased ($P < 0.05$ or $P < 0.01$) in all of the PFOS dose groups. Additionally, peroxisomal acyl-CoA oxidase, CYP 4A14, and acyl-CoA dehydrogenase mRNA transcripts were increased in the 5 and 10 mg/kg-d dose groups, observations that suggested degradation of long-chain fatty acids by peroxisomes. Increases in peroxisomal oxidation, in the absence of increased mitochondrial β-oxidation, has the potential to promote the accumulation of fatty acids in the liver (i.e., steatosis).

The LOAEL identified for this study is 5 mg/kg-d. Furthermore, at 1 mg/kg-d, increased liver weights were noted in the absence of histopathological correlates. Thus, the 1 mg/kg-d dose was determined to be a NOAEL (Wan *et al.* 2012). It was concluded that the hepatic changes observed in this mouse model were similar to those associated with non-alcoholic fatty liver disease in humans and did not necessarily reflect peroxisome proliferator-activated receptor alpha (PPARα) activation in their entirety.

A relatively recent study explored the acute (0–14 day assessment following exposure) and sub-acute (0–30 day assessment following exposure) toxicological outcomes in response to dose-dependent administration of PFOS by oral gavage to six-week-old male C57BL/6J mice (Xing *et al.* 2006). This group showed that, in the acute toxicity component of the study, the acute oral LD_{50} of PFOS in male C57BL/6J mice was 0.579 g/kg BW. Furthermore, in the sub-acute study component, the oral toxicity of PFOS at 2.5, 5, and 10 mg PFOS/kg BW/d disrupted antioxidative homeostasis, induced hepatocellular apoptosis, and induced liver injury (as shown by increased serological levels of the liver enzymes aspartate transaminase (AST), alanine transaminase (ALT), alkaline phosphatase (ALP), and gamma glutamyl transpeptidase (gamma-GT) and by morphological changes as revealed by histology). The authors concluded that exposure to PFOS increased the liver size and relative liver weight of the mice. PFOS treatment also caused liver damage but only marginally affected functional features of the kidneys and spleen in this mouse model (Xing *et al.* 2016).

In another study, groups of 10 male 3-month-old SD rats/group were administered PFOS in purified Milli-Q water at a dose of 5 or 20 mg/kg-d by oral gavage (intragastric intubation) for 28 days (Cui *et al.* 2009). Rats were sacrificed following the PFOS-exposure period, and blood and tissue samples were obtained. All rats that received a dose of 20 mg/kg-d PFOS experienced significant loss in body weight and none survived beyond day 26 post-exposure. Rats exposed to 5 or 20 mg/

kg-d showed consistent hepatocellular hypertrophy, cytoplasmic vacuolation, and focal necrosis on histopathological examination. However, the effects were more pronounced in the 20 mg/kg-d-treated group, and included focal hemorrhage, erythrocytic transudation, and focal hepatic degeneration, accompanied by inflammatory cell infiltrates. These observations confirmed that the liver was a target organ of the sub-acute effects of PFOS (Cui *et al.* 2009). The lung was also a target organ of PFOS-mediated toxicity.

At a dose of 5 mg/kg-d, pulmonary congestion and focal or diffuse thickened epithelial walls were seen, which were more apparent at the higher 20 mg/kg-d dose of PFOS, wherein rats displayed swelling and discoloration of the liver, with hepatocyte hypertrophy and cytoplasmic vacuolation observed following histopathological examination. Viscera indices were calculated including the hepatosomatic index (HSI), renal-somatic index (RSI), and gonad-somatic index (GSI) to evaluate the hyperplasia, swelling, and/or atrophy of the organs, and all three indices were statistically significantly increased in all of the treated groups, relative to the control.

This group hypothesized that insufficient production of pulmonary surfactant by alveolar type II cells was the major cause of death and severe pulmonary injury in the rats treated with 20 mg/kg-d PFOS (Cui *et al.* 2009). From this 28-day study, the LOAEL was determined to be 5 mg/kg-d in rats, which was based on a significant decrease in body weight, dose-related effects in the liver, and pulmonary congestion. Although a NOAEL could not be derived, other research groups determined a NOAEL of 0.025 mg/kg for PFOS (OECD, 2002). Table 3.1 summarizes the acute and sub-acute toxicological effects of orally administered PFOS.

Mammalian Oral Toxicity – Sub-Chronic

The effects of PFOS exposure were determined in two related studies conducted in rhesus monkeys (Goldenthal *et al.* 1978, 1979). A summary of the sub-chronic oral toxicity in mammals is presented in Table 3.2.

In the first study, two monkeys/sex/dose group were treated with PFOS dissolved in distilled water by oral gavage at doses of 0, 10, 30, 100, or 300 mg/kg-d. This study was terminated on day 20 as all of the treated monkeys in the 300 mg/kg group had failed to survive PFOS exposure, beginning on day 4 following treatment. Deaths were also observed at all lower doses, although the authors did not confirm whether this occurred in both or only one sex of the animals. In all groups, clinical signs of toxicity were observed and included decreased activity, emesis, body stiffening, general trembling, twitching, weakness, and convulsions. At necropsy, several of the monkeys in the high-dose groups (i.e., 100 and 300 mg/kg-d) showed yellowish-brown discoloration of the liver with no evidence of visible microscopic lesions. Calculated NOAEL or LOAEL values were not determined in this study (Goldenthal *et al.* 1978).

In the second study, two rhesus monkeys/sex/dose were treated with PFOS in distilled water at doses of 0, 0.5, 1.5, or 4.5 mg/kg-d in a 90-day study (Goldenthal *et al.* 1978, 1979). All monkeys in the 4.5 mg/kg-d-dosed group died or were euthanized by week 7 and displayed decreased body weight gain, signs of severe gastrointestinal tract toxicity (i.e., including emesis, black stools, and anorexia), decreased

TABLE 3.2
Sub-Chronic Oral Toxicity in Mammals

Test Organism	Test Duration	Test Results			Reference
		NOAEL (mg/kg-d)	LOAEL (mg/kg-d)	Effects Observed at the LOAEL	
Mouse	60 days	0.008	0.083 Males 0.833	Immunotoxicity study. Derived from increases in splenic natural killer (NK) cell activity and increases in liver organ weight Based on decreased NK cell activity	Dong et al. 2009
Mouse	90 days	0.43	2.15	Neurotoxicity study. Based on changes in water maze activities and histopathology examination of hippocampus	Long et al. 2018
Rat	90 days (oral, i.e., supplied in diet)	NA	2.0	Derived from increased liver organ weight, hepatocyte hypertrophy, and decreased food consumption	Goldenthal et al. 1978
Rat	Females were dosed via oral gavage for 42 days prior to mating, during mating (for and with untreated breeder males), and subsequently through gestation, parturition and lactation (females) across two generations.	0.10 F₀ (male and female) parents 0.4 F₁ (male & female) parents 0.4 F₁ offspring 0.1 F₂ offspring	0.40 F₀ (male and female) parents NA (male & female) parents 1.6 F₁ offspring 0.4 F₂ offspring	Reproductive study. Derived from observations of decreased adult body weight gains and decreased food consumption. Higher dose not tested Based on decreased viability of offspring and decreased body weight. Based on decreased body weight	Luebker et al. 2005b

(Continued)

TABLE 3.2 (CONTINUED)
Sub-Chronic Oral Toxicity in Mammals

Test Organism	Test Duration	NOAEL (mg/kg-d)	LOAEL (mg/kg-d)	Test Results Effects Observed at the LOAEL	Reference
Rat	Six weeks prior to mating and continued through mating, gestation, and lactation day (LD 4) (63 days)	0.4 F0 dams NA F1 offspring	0.8 F0 dams 0.4 F1 offspring	Based on decreased body weight gains. Based on decreased body weight of pups BMDL5 values (i.e., based on the lower 95% confidence limits of the BMD5) were estimated for decreased gestation duration (0.31 mg/kg-d) and viability of the pups (0.89 mg/kg-d)	Luebker et al. 2005a
Monkey	90 days	NA	NA	Derived from observations of diarrhea and anorexia in treated animals	Goldenthal et al. 1979

Key: NA = not applicable; GD = gestation day; LD = Lactation day activity, marked/severe rigidity, and reduced concentrations of serum cholesterol. Additionally, histopathological analysis of the monkeys that were dosed at 4.5 mg/kg-d showed diffuse lipid depletion in the adrenals (4/4), diffuse atrophy of the pancreatic exocrine cells (3/4), and moderate diffuse atrophy of the serous alveolar cells (3/4). All monkeys in the 0.5 and 1.5 mg/kg-d groups survived; however, these monkeys exhibited dose-related increases in clinical signs that included occasional diarrhea, soft stools, and anorexia. Evidence of low levels of serum cholesterol in the 1.5 mg/kg-d group (1/4) was also seen. Body weight decreases were also seen in males and females of the 1.5 mg/kg-d group. Treatment-related effects were not seen for either the 0.5 or 1.5 mg/kg-d groups at the time of necropsy. Based on the findings from these studies, the LOAEL was 0.5 mg/kg-d (with a feed LOAEL of 10 mg/kg-d), although the NOAEL could not be determined from these data (Goldenthal *et al.* 1978, 1979).

In a similar study (Seacat *et al.* 2002), of crab-eating macaques (*Macaca fascicularis*), the sub-chronic toxicity of K+PFOS was explored. PFOS was administered orally in a capsule by intragastic intubation daily for 26 weeks (or 182 days) across the dose range of 0, 0.03, 0.15, or 0.75 mg/kg-d. In this work, six young-adult to adult macaques/sex/group, except for the 0.03 mg/kg-d group (four monkeys/sex) were exposed to PFOS. Two monkeys per sex in the control, 0.15, and 0.75 mg/kg-d groups were monitored for 1 year (a 52-week recovery period) post-exposure to detect reversible or delayed toxic effects from the potassium salt of PFOS. Monkeys were observed twice daily for mortality, morbidity, clinical signs, and qualitative food intake. Body weights were recorded pre-dosing and weekly thereafter, and ophthalmic examinations were performed pre- and post-treatment. At the time of sacrifice, PFOS levels were assayed in serum and liver tissue, and hematology and clinical chemistry were performed. Urine and fecal analyses were carried out, with full histopathology being performed at the scheduled sacrifice. Liver samples were also obtained for hepatic peroxisome proliferation, and immunohistochemistry was performed by the proliferating cell nuclear antigen assay (PCNA) to study cell proliferation (Seacat *et al.* 2002).

Serum PFOS measurements showed a linear increase with dosing duration in the 0.03 and 0.15 mg/kg-d groups and a non-linear increase in the 0.75 mg/kg-d group. PFOS levels in the high-dose group appeared to plateau after about 100 days post-exposure. Serum PFOS levels decreased during recovery in the two highest-dosed groups. The average percentage of the cumulative PFOS dose in the liver at the end of the treatment ranged from 4.4% to 8.7%, with no difference among dosed groups or between genders.

The concentration of PFOS in the liver decreased during the recovery period (Seacat *et al.* 2002). Additionally, significant adverse effects occurred only in the 0.75 mg/kg-d group. In this particular group, two male monkeys died. One died on day 155 and one monkey was found moribund and was sacrificed on day 179. The monkey that did not survive treatment exhibited pulmonary necrosis and severe acute recurrences of pulmonary inflammation as its cause of death. The specific cause of the moribund condition, although unclear, was suggested to be hyperkalemia by clinical chemistry analyses. Overall gain in mean body weight was less in the

0.75 mg/kg-d for males and females after PFOS treatment when compared with controls ($P \leq 0.05$). Also, mean absolute and relative (to body weight) liver weights were markedly increased in the 0.75 mg/kg-d for the male and female monkey groups.

Administration of doses of up to 0.75 mg/kg-d PFOS (potassium salt) to monkeys for 26 weeks did not cause any significant gross or microscopic alterations in the heart or aorta (Seacat *et al.* 2002). In addition, male and female monkeys in the 0.75 mg/kg-d dosing group had lower total serum cholesterol concentration, beginning on day 91 (27%–68% [males] and 33%–49% [females] lower than the controls) and lower high-density lipoprotein (HDL) cholesterol concentration beginning on day 153 (i.e., 72%–79% and 61%–68% lower in males and females, respectively, than in the controls) when compared with the control values. However, this effect was reversible as the total cholesterol concentrations were similar to those in the controls by week 5 during recovery, and the total high-density lipoprotein cholesterol was similar to controls by week 9 during recovery. Estradiol concentrations were lower in the 0.75 mg/kg-d dosed males and females on day 182; however, the data were highly variable and it was stated by the authors that the changes in observed estradiol levels in the 0.75 mg/kg-d dosed groups was poorly understood (Seacat *et al.* 2002). Authors hypothesized that the lowered estradiol was mediated by a feed-back loop in an attempt to sustain testosterone levels in males, e.g., decreased levels in, or direct blocking of aromatase activity. By contrast, the relatively low estradiol levels that were seen in female macaques might have been proportional to lower levels of High Density Lipoprotein (HDL). Total triiodothyronine (T3) concentrations were decreased and thyroid-stimulation hormone (TSH) concentrations were increased on day 182 in the high-PFOS dosed monkeys. However, a true dose-response was absent because there was no evidence of clinical hypothyroidism; since the TSH values were within the reference range, there was no evidence of hyperlipidemia, and thyroid gland histopathological lesions were absent.

Hepatic peroxisome proliferation was measured in terms of hepatic palmitoyl coenzyme A oxidase (PCoAO) activity and was markedly higher in the 0.75 mg/kg-d females than in the controls, although the increase was not dose-related and was less than two-fold. There were no treatment-related effects on cell proliferation in the liver, pancreas, or testes when assayed by the proliferating cell nuclear antigen immunohistochemistry cell-labeling index. Two high-dose males and one high-dose female had mottled livers on gross examination at sacrifice; this was also observed in the high-dose male that died during the study. All females and 3/4 males at the high-dose treatment had centrilobular or diffuse hepatocellular hypertrophy (Seacat *et al.* 2002).

Based on observations of decreased body weight gain, decreased serum cholesterol concentration, lowered triiodothyronine and estradiol concentrations, as well as observations of increased liver weights and histopathological lesions in the liver, the calculated NOAEL in male and female macaques was 0.15 mg/kg-d in which the serum PFOS concentration was 83 µg/mL in males and 67 µg/mL in females. By contrast, the LOAEL in male and female macaques was 0.75 mg/kg-d, in which the serum concentration of PFOS was 173 µg/mL in males and 171 µg/mL in females, which represents an approximately two-fold difference compared with those for the

NOAEL for serum levels of PFOS, and this was despite a 5-fold difference in the administered dose of PFOS (Seacat *et al.* 2002).

A sub-chronic toxicity 90-day feed study was conducted in SD rats (five rats/ sex/dose), using feed that contained PFOS at 0, 30, 100, 300, 1,000, or 3,000 ppm (Goldenthal *et al.* 1978). The dietary levels were thus equivalent to 0, 2, 6, 18, 60, or 200 mg/kg-d, respectively. All rats in the 300 ppm (18 mg/kg-d) group exhibited emaciation, convulsions, increased sensitivity to external stimuli, hunched-back posture, and reduced activity, and ultimately began dying on day 7. At 100 ppm (6 mg/kg-d), body weights had decreased (i.e., by approximately 16.5%) and the liver weights had increased. Furthermore, in the 100 ppm group, increased liver enzyme activities, hepatic vacuolization, hepatocellular hypertrophy, and weight loss were observed, and, ultimately, three male and two female rats died.

All treated rats displayed very slight to slight cytoplasmic hepatocyte hypertrophy in the liver. At 30 ppm (2 mg/kg-d), rat mean body weights were slightly lower and food consumption was lower (male rats), whereas increased absolute and relative liver (female rats) weights were observed. All rats survived in the 30 ppm-dosed group. Based on observations of significant decreases in food consumption and increases in absolute and relative liver weight, the LOAEL for PFOS was 30 ppm (equivalent to 1.5–2.0 mg/kg-d) but the NOAEL could not be determined from the available data. Compared to the controls, slight elevation of blood glucose concentrations, serum alkaline phosphatase activity, blood urea nitrogen (BUN) concentration, and gamma glutamyltransferase activity were also reported (Goldenthal *et al.* 1978).

Similarly, other research groups had performed an interim sacrifice in a sub-chronic study that explored the dietary toxicity of potassium PFOS that was conducted on Sprague-Dawley Crl:CD (SD) IGS BR rats (Seacat *et al.* 2003). Animals were exposed to 0, 0.5, 2.0, 5.0, or 20 ppm/rat/sex/dose in a study that was derived from a 2-year chronic cancer bioassay. Data that were reported here were derived from interim sacrifices that were made at 4 or 14 weeks following PFOS exposure (Seacat *et al.* 2003). At 4 weeks, doses of PFOS were equivalent to 0, 0.05, 0.18, 0.37, or 1.51 mg/kg-d in males and 0, 0.05, 0.22, 0.47, or 1.77 mg/kg-d in females, respectively, at the time of sacrifice.

At 14 weeks, PFOS doses in the diet were equivalent to 0, 0.03, 0.13, 0.34, or 1.33 mg/kg-d in male rats and 0, 0.04, 0.15, 0.40, or 1.56 mg/kg-d in female rats, respectively. Observations mostly identified dose-dependent effects following PFOS exposure. Animals were observed twice daily for evidence of mortality and morbidity, with a weekly clinical exam of the animals also included. Food use efficiency was determined, and mean daily PFOS intake, cumulative dose, and percentage of dose were identified in the liver and sera. Blood and urine were obtained from 10 animals/sex/dose during week 4 for clinical chemistry, hematology, and urinalysis evaluation. A thorough necropsy was performed on five animals/sex/dose at the end of 4 weeks of treatment. At that time point, it was determined that the effects of PFOS exposure included decreased serum glucose and a near two-fold increase in hepatic palmitoyl coenzyme A oxidase activity (a surrogate measure for peroxisome proliferation) in male rats dosed at 20 ppm, whereas, in female rats, increased activities of ALT were observed. At 14 weeks, increased liver weights, decreased serum cholesterol concentration, and increased ALT activity was seen in the 20 ppm-dosed male rats, although elevated hepatic palmitoyl co-enzyme A oxidase activity was not sustained at 14 weeks. In addition, relative liver weights and BUN levels were increased in male and female rats at 14 weeks. Hepatocytic hypertrophy and cyto-plasmic vacuolation were seen in the 5 or 20 ppm male and the 20 ppm female dose groups. Serological PFOS levels, although generally higher in female than male rats, were proportional to both dose and cumulative dose (Seacat *et al.* 2003).

No effects were seen on body weight, food use efficiency, urinalysis evaluation, or peroxisome proliferation (hepatic PCoAO activity was unchanged) at 14 weeks. All significant changes, when compared with controls, were seen in the highest dose group. Food consumption was decreased. Absolute liver weights increased signifi-cantly in males and relative (to body weight) liver weights increased in both males and females, respectively. All hematology parameters were similar to those in the controls. Clinical chemistry parameters that were affected, compared to controls, included decreased serum cholesterol concentration (males), increased ALT activity (males), and increased urea nitrogen (males and females). From these observations, the calculated LOAEL for males and females which had been administered PFOS in the diet up to 14 weeks was 20 ppm (i.e., corresponding to 1.33 mg/kg-d in males and 1.56 mg/kg-d in females) and the calculated NOAEL was 5 ppm (i.e., corresponding to 0.34 mg/kg-d in males and 0.4 mg/kg-d in females) (Seacat *et al.* 2003). A sum-mary of the sub-chronic oral toxicity results of PFOS in mammals is presented in Table 3.2.

MAMMALIAN ORAL TOXICITY: DEVELOPMENTAL/REPRODUCTIVE EFFECTS

Previous studies have confirmed that rats and mice exposed to orally administered PFOS were affected in developmental and reproductive studies. Prenatal exposure of rats to PFOS increased neonatal mortality when dams were given PFOS doses ≥ 1 mg/kg-d, and decreased pup body weight was seen at maternal doses of 0.4 mg/kg-d. Neonatal death was demonstrated to be a direct consequence of PFOS on the pul-monary surfactant. Other developmental and reproductive toxicity effects included decreased gestation length and delays in development. Higher doses of PFOS expo-sure resulted in fetal sternal defects and cleft palate in both rats and mice. In addi-tion, many specialized developmental studies have also been conducted with PFOS to assess in greater detail the long-term effects of PFOS exposure on offspring. Postnatal effects of gestational and lactational exposure included evidence that sup-ported developmental neurotoxicity, changes in thyroid and reproductive hormones, altered lipid and glucose metabolism, and dampened functional immunocompetence (DeWitt *et al.* 2012).

Developmental toxicity studies were conducted with PFOS and 2-*N*-ethylperfluorooctane sulfonamido ethyl alcohol (*N*-EtFOSE) in rats and rabbits that were orally (by intubation) administered those test articles during the gestation period at doses of 0, 1, 5, 10, or 20 mg/kg-d in rats for the *N*-EtFOSE study, and 0, 0.1, 1, 2.5, or 3.75 mg/kg-d in mated rabbits on days 7 through to 20 of gesta-tion for the PFOS and *N*-EtFOSE study (Case *et al.* 2001). It was concluded that no compound-related clinical signs or deaths occurred in the developmental toxic-ity studies of the dosed pregnant female animals. However, maternal toxicity was observed at the higher doses, as shown by reduced body weight gains and feed con-sumption in three studies. No compound-related malformations were observed in either dosed rats or rabbits. In the rabbit component of the study and maternal weight gain, an effect concentration NOAEL of 0.1 mg/kg-d was determined. It was con-cluded that these compounds were not selective developmental toxicants in either rats or rabbits in this study (Case $et\ al.$ 2001).

Thibodeaux $et\ al.$ 2003 studied the maternal and developmental toxicities of PFOS in SD rats and CD-1 mice. Groups of 9–16 pregnant rats were given PFOS in 0.5% Tween-20 at daily doses of 1, 2, 3, 5, or 10 mg/kg-d by gavage from ges-tational day (GD) 2 through to GD 20. Rats were euthanized on GD 21. Maternal body weights and food and water consumption decreased in a dose-dependent manner at doses greater than 2 mg/kg-d. A dose-dependent increase in the serum PFOS concentration was observed, with liver concentrations that were approxi-mately four times higher than was found in serum at each dose. Liver weight was unaffected in the treated rats. Serum chemistry showed decreases in cholesterol (14% lower than in the controls) and triglyceride concentrations (34% lower than in the controls) at 10 mg/kg-d. Serum levels of the thyroid hormones thyroxine (T4) and triiodothyronine (T3) were decreased in all treated rats when compared with controls, although a feedback response on the thyroid-stimulating hormone (TSH) was not observed. In rats, PFOS did not alter the number of implantations or live fetuses at term, although small decreases in fetal weights were noted. Additionally, significant increases in the incidence of developmental effects were noted and included presence of cleft palate, defective sternebrae, anasarca, enlarged right atrium, and ventricular septal defects, which were particularly noted in the rats receiving the highest dose of PFOS (10 mg/kg-d).

Developmental neurotoxicity studies of PFOS have also been informative (Butenhoff $et\ al.$ 2009, Wang $et\ al.$ 2015, Zeng $et\ al.$ 2011). In the study by Butenhoff $et\ al.$ (2009), 25 female SD rats/group were exposed to potassium PFOS at 0, 0.1, 0.3, or 1.0 mg/kg-d by gavage from GD 0 through to postnatal days (PND) 20 (Butenhoff $et\ al.$ 2009). An additional ten mated females/group were used to collect additional blood and tissue specimens. Offspring were monitored through PND72 for growth, maturation, motor activity, learning and memory, acoustic startle reflex, and brain weight.

Treatment-related effects were not seen in the context of pregnancy rates, gesta-tion length, number of implantation sites, number of pups born, sex ratio of pups, measures of pup survival, pup body weights, and gross anatomical findings. Maternal body weight and body weight gain during gestation were comparable between the treated and control groups. On lactation days (LDs) 1–4, dams in the 1.0 mg/kg-d group had slight, but not significantly, lower weight gain and food consumption than did their control counterparts, which resulted in lower ($P < 0.05$ or 0.01) absolute body weight throughout lactation. Food consumption was transiently decreased on GD 6–9 for the 0.3 mg/kg-d group and on GD 6–12 for the 1.0 mg/kg-d group. In treated dams, these findings are not considered to be treatment-related or indeed adverse. From the collated results, the maternal toxicity NOAEL was 1.0 mg/kg-d and the LOAEL could not be determined (Butenhoff *et al.* 2009).

Treatment-related effects were not observed on functional observational battery assessments performed at PNDs 4, 11, 21, 35, 45, and 60. Male offspring from dams that were administered 0.3 and 1.0 mg/kg-d PFOS had statistically significant ($P < 0.05$) increased motor activity at PND 17, compared with the control, but this was not observed at PND 13, 21, or 61. No effect on habituation was observed in the 0.1 and 0.3 mg/kg-d males or in the 1.0 mg/kg-d females. At PND 17, males at 1.0 mg/kg-d showed a lack of habituation, as illustrated by significantly ($P < 0.05$) increased activity counts for the sequential time intervals of 16–30, 31–45, and 46–60 minutes. The normal habituation response is for motor activity to be highest when the animals are first exposed to a new environment and to decline during subsequent exposures to the same environment, as they have learned what to expect. There were no effects of PFOS exposure on acoustic startle reflexes or in the Biel swimming maze trials in males or females. Mean absolute and relative (to body weight) brain weight and brain measurements (length, width) were similar between the control and the treated animals. Based on the increased motor activity observed, reflecting decreased habituation, the LOAEL for developmental neurotoxicity in male rats was 1.0 mg/kg-d and the NOAEL was 0.3 mg/kg-d (Butenhoff *et al.* 2009).

The effects of PFOS on spatial learning and memory following pre- or post-natal exposure to PFOS were also investigated (Wang *et al.* 2015). PFOS was administered to pregnant Wistar rats in the drinking water at 0, 5, or 15 mg/L, beginning on GD 1 and continuing through lactation. However, doses to the animals were not calculated, and body weight and water consumption data were not presented. Doses were estimated as 0, 0.8, or 2.4 mg/kg-d, which were derived using sub-chronic values for female Wistar rats from the U.S. EPA (1998) protocol. At PND 7, maternal serum PFOS levels in the treated (0.8 and 2.4 mg/kg-d) groups were 25.7 and 99.3 µg/mL, and at PND 35, maternal serum PFOS levels in the treated groups were 64.3 and 207.7 µg/mL, respectively.

At PND 1, pups were cross fostered to establish groups for untreated controls, prenatal exposure only, post-natal exposure only, and continuous exposure. After weaning, pups were given the same treated or control water as their foster dam. Three pups per group were sacrificed on PNDs 7 and 35 to quantify hippocampal protein and mRNA levels. On PND 35, 8–10 pups per group were tested in the Morris water maze, which consisted of one day of visible platform tests, seven days of hidden platform tests, and a probe trial 24 hours after the last hidden platform test. Offspring survival from high-dose dams on PND 1 was reduced before cross fostering; survival data on PND 5 were not presented. On water maze testing day 1, swimming speed and the time to reach the visible platform were similar between all treated and control groups.

Escape latency was increased for all treated groups on one or more testing days. The most pronounced and significant effect was in pups exposed prenatally from dams given 15 mg/L and cross fostered to the control dams. Similar trends were observed for escape distance. During the probe trial for memory testing, pups continuously exposed pre- and post- natally to 15 mg/L dosage spent less time in the target quadrant than did the unexposed controls but statistical significance was not achieved as consistently as that for the group exposed only during gestation. Abundance of growth-associated protein-43, neural cell adhesion molecule 1, nerve growth factor, and brain-derived neurotrophic factor were decreased in the hippocampus on PND 35, especially in pups exposed prenatally to 15 mg/L PFOS and cross fostered to the control dams.

In a separate but related study, ten pregnant SD rats per group were administered PFOS in 0.5% Tween 80 at doses of 0, 0.1, 0.6, or 2.0 mg/kg-d by oral gavage from GD 2 through to GD 21 (Zeng *et al.* 2011). At GD 21, onset of parturition was monitored in the dams, and the day of delivery was referred to as PND 0, at which time five pups/litter were sacrificed and the trunk blood, cortex, and hippocampus harvested for further study. Morphological cellular changes were associated with expression of astrocyte activation markers, glial fibrillary acidic protein (GFAP), and S100 calcium-binding protein B, as shown by immunohistochemistry. Remaining pups were randomly allocated to dams from the different dosage groups and permitted to nurse up to PND 21, at which time pups were sacrificed and tissues harvested as described above for PND 0. A dose-dependent increase in PFOS levels was found in the serum, whereas the levels in the hippocampus and cortex tended to be lower in all tissues at postnatal day (PND) 21 as compared with PMD 0. Inflammatory responses included increased hippocampal mRNA expression of the interleukin IL-1β and the tumor necrosis factor TNFα at PMD 0 in all treated groups as compared with controls, and in those from dams administered $\geq$ 0.6 mg/kg-d at PND 21. In the cortex, IL-1β and TNF-α expression were only significantly increased in the 0.6 mg/kg-d group and 2.0 mg/kg-d group, respectively, at PND 0. At PND 21 in the cortex, IL-1β was increased at $\geq$ 0.6 mg/kg-d, and TNF- α was increased in the high-dose group (Zeng *et al.* 2011).

To determine the mechanisms of PFOS-mediated inflammation, the mRNA levels of activation protein-1 (AP-1), nuclear factor-κB (NF-κB), and cAMP response element-binding protein in the hippocampus and cortex were determined. The hippocampus displayed the highest increases in expression of AP-1 in all dose groups, and an increase in both NF-κB and cAMP response element-binding protein for the $\geq$ 0.6 mg/kg-d groups at PND 0. The expression of two synaptic proteins, synapsin 1 (Syn 1) and synaptophysin (Syp), decreased in the hippocampus. However, Syp expression was increased in the cortex (Zeng *et al.* 2011).

Lau *et al.* (2003) studied the post-natal effects in rats and mice following *in utero* exposure to PFOS. In this work, pregnant SD rats were exposed to PFOS at doses of 1, 2, 3, 5, or 10 mg/kg-d by oral gavage from GD 2 through to GD 21. Similarly, pregnant CD-1 mice were exposed to PFOS at doses of 1, 5, 10, 15, or 20 mg/kg-d from GD 1 through to GD 17. At parturition, all animals were initially born alive and were seen to be active. However, within 30 to 60 min post-parturition, the neonatal pups were pale, inactive, and moribund in the highest-dosage groups (10 mg/kg-d for rat and 20 mg/kg-d for mouse). All animals died soon after these initial observations. In addition, in the groups treated with the second-highest doses, namely 5 mg/kg-d (rat) and 15 mg/kg-d (mouse), the neonates also became moribund but survived for a longer period of time (8–12 hours). Over 95% of these animals died within 24 hours. Approximately 50% of offspring died at the 3 mg/kg-d dose for rats and in the 10 mg/kg-d dose for mice. Survival of pups in the 1 and 5 mg/kg-d treated dams was similar to that in the controls. Additionally, cross fostering the PFOS-exposed rat neonates (5 mg/kg-d) to the control group nursing dams did not improve survival rates (Lau *et al.* 2003).

Increases in liver weights were observed in PFOS-exposed mice at all dose levels at different post-natal days. The serum thyroxine levels were suppressed in PFOS-exposed rat pups, although the concentrations of triiodothyronine and thyroid-stimulating hormone (TSH) levels were unchanged. In addition, developmental learning, as tested by T-maze delayed alternations in weanling rats, was also unaffected. Observations suggest that *in utero* exposure to PFOS adversely affected post-natal survival of neonatal rats and mice, and retarded both neonatal growth and development – an observation that was accompanied by hypothyroxinemia in the surviving rat pups (Lau *et al.* 2003). No dose- or treatment-related effects were observed on T4, T3, and TSH levels in the pups. The LOAEL for this study in mice was 5 mg/kg-d and the NOAEL was 1 mg/kg-d. The authors calculated a $BLDL_5$ for survival at 6 days of 3.88 mg/kg-d (Lau *et al.* 2003).

Luebker *et al.* (2005a) studied two-generation reproductive and cross-fostering effects of PFOS in male and female SD rats, that were dosed orally at 0, 0.1, 0.4, 1.6, or 3.2 mg/kg-d prior to mating, during mating, and, for female rats, through gestation and lactation, across two successive generations. Due to marked F_1 neonatal toxicity that was observed in the 1.6 and 3.2 mg/kg-d groups, continuation of the F_1 progeny into the second generation was limited to F_1 pups from the 0, 0.1, and 0.4 mg/kg-d groups. PFOS did not affect reproductive performance, mating, or fertility; however, reproductive outcome, as demonstrated by decreased length of gestation, number of implantation sites, and increased numbers of dams with stillborn pups or with all pups dying on LD 1–4, was affected at 3.2 mg/kg-d in F_0 dams. These effects were not observed in F_1 dams at the highest dose tested, 0.4 mg/kg-d. Neonatal toxicity in F_1 pups, as demonstrated by reduced survival and body weight gain through to the end of lactation, occurred at a maternal dose of 1.6 mg/kg-d and higher, but was not seen at 0.1 or 0.4 mg/kg-d or in F_2 pups at 0.1 or 0.4 mg/kg-d. From these data, the NOAELs for reproductive function were $F_0 \geq 3.2$ and $F_1 \geq 0.4$ mg/kg-d, and for reproductive outcome were $F_0 = 1.6$ and $F_1 \geq 0.4$ mg/kg-d. This study suggested that PFOS exposure *in utero* contributed to post-natal pup mortality, and that pre- and post-natal PFOS exposure had additive effects with respect to the toxic impacts observed in pups (Leubker *et al.* 2005).

In complementary work by the same group (Leubker *et al.* 2005), a sub-chronic SD rat gavage study of gestation length and pup viability was designed with the aim of better defining the dose-response curve over the selected PFOS doses of 0.8, 1.0, 1.2, and 2 mg/kg-d for neonatal mortality in rat pups born to PFOS-exposed dams (Luebker *et al.* 2005). A decreased viability trough at lactation day (LD) 5 was observed in the 0.8 mg/kg-d group and higher. Reduced neonatal survival did not appear to dampen the levels of serum lipids, glucose, or thyroid hormones. The end points of gestation length and decreased viability were positively correlated, which suggested that late-stage fetal development might be adversely affected following exposure to PFOS *in utero*, which might also contribute to overall mortality.

This study derived a NOAEL of 0.37 mg/kg BW-d and a NOAEC of 7.4 mg/kg feed (Leubker *et al.* 2005). In addition, benchmark dose (BMD) estimates for decreased gestation length, birth weight, pup weight on LD 5, pup weight gain through LD 5, and viability gave values ranging from 0.27 to 0.89 mg/kg-d for the lower 95% confidence limit of the BMD_5 ($BMDL_5$).

In one study, CD-1 mice were treated in groups of 20–29 CD-1 mice and administered 0, 1, 5, 10, 15, or 20 mg/kg-d PFOS at GD 1–17 (Thibodeaux *et al.* 2003). Maternal weight gain, food and water consumption, and serum clinical chemistries were monitored and recorded. Mice were euthanized on GD 18. Measured parameters, as described for the rat, were also measured in the mice. Control mice received 0.5% Tween-20 as the vehicle. Exposure of pregnant mice to PFOS throughout gestation had no effect on implantation site numbers; however, a marked increase in post-implantation loss was seen in the 20 mg/kg-d dose group (Thibodeaux *et al.* 2003). In PFOS-exposed mice, investigators noted deficits in maternal weight gains, although these were not as pronounced as was seen in the rat (Thibodeaux *et al.* 2003). Maternal body weight gain was significantly decreased at 20 mg/kg-d. Food and water consumption were not affected by PFOS treatment. Increases in serum PFOS were comparable to those in the rat. PFOS treatment increased ($P < 0.05$) the liver weight in a dose-dependent manner in the mice. T4 was decreased in a dose-dependent manner at GD 6 with statistical significance ($P < 0.05$) attained for the 20 mg/kg-d group; levels of T3 and TSH were not affected by PFOS treatment. A significant increase in post-implantation loss was seen in animals administered 20 mg/kg-d and reduced fetal weight ($P < 0.05$) was observed from dams in the 10 and 15 mg PFOS/kgd groups. Birth defects, such as cleft palate, ventricular septal defect, and right atrium enlargement were seen at doses ≥ 10 mg/kg-d.

In another study, ten pregnant ICR mice/group were exposed to 0, 1, 10, or 20 mg/kg-d of PFOS daily by gavage at GD 1 to GD 17 or 18 (Yahia *et al.* 2008). Five dams/group were sacrificed on GD 18 for fetal external, skeletal effects, and histology of the maternal liver, kidneys, lungs, and brain; the other five were left to give birth. Body weight, food consumption, and water consumption were monitored in the dams. In the dams sacrificed at GD 18, the gravid uterus was removed, and the number of live/dead fetuses, fetal body weight, and number of resorptions were all recorded. Four pups/litter were sacrificed immediately after birth to examine their lungs. All dams survived and without evidence of any clinical signs. A marked decrease in body weight was observed in the dams administered 20 mg/kg-d PFOS, beginning at GD 10. Water consumption was increased. Maternal absolute liver weight increased in a dose-dependent and significant manner in the 10 (59% increase in absolute liver weight) and 20 (60% increase) mg/kg-d groups.

All neonates in the 20 mg/kg-d dose group were born pale, weak, and inactive, and all died within a few hours of birth. At 10 mg/kg-d, 45% of those born died within 24 hours, whereas survival of the 1 mg/kg-d group was similar to that of the controls. Neonatal weight was lower at 10 and 20 mg/kg-d than in the controls. In the fetuses from dams treated with 20 mg/kg-d, there was a very high incidence of cleft palates (98.56%), sternal defect (100%), delayed ossification of phalanges (57.23%), wavy ribs (84.09%), spina bifida occulta (100%), and curved fetus (68.47%). Similar defects were seen in fetuses from dams treated with 10 mg/kg-d, except at a lower incidence (Yahia *et al.* 2008). Histology showed that all fetuses examined at GD 18 from dams treated with 20 mg/kg-d were alive and had normal lung structures but mild to severe intracranial dilatation of the blood vessels. Neonates from the 20 mg/kg-d-treated dams had fetal lung atelectasis (partial or complete collapse of the lung or a lobe of the lung), with reduced alveolar space and intracranial blood vessel dilatation when examined histopathologically. Three neonates from each of the five dams treated with 10 mg/kg-d were examined, and 27% had slight lung atelectasis and 87% had mild to severe dilatation of brain blood vessels. Based on the significant increase in liver organ weight, the maternal LOAEL was 10 mg/kg-d and the NOAEL was 1 mg/kg-d. Based on the abnormalities observed in the fetuses and the decreased survival rate, the developmental LOAEL was 10 mg/kg-d and the NOAEL was 1 mg/kg-d (Yahia *et al.* 2008).

Another study treated 8–10 pregnant CD-1 (Charles River) mice/group with 0 to 6 mg/kg-d PFOS in 0.5% Tween-20 by daily gavage from GD 12 to GD18 (Fuentes *et al.* 2007). Following treatment, mice were either left undisturbed or immobilized three times per day for 20 min to induce maternal stress. Maternal body weight and consumption of food and water were then monitored. The length of gestation, number of live/dead pups, and sex/weight of pups were then recorded post-partum. Body weights of the pups were recorded, landmarks for development were monitored, and neuromotor maturation tests (i.e., surface righting reflex, forelimb grip strength) were conducted during the post-natal phase of development (Fuentes *et al.* 2007). When the pups were aged three months, development was assessed by open-field and rotarod tests. It was determined that PFOS exposure did not impact maternal body weight or food/water consumption. At PNDs 4 and 8, reduced body weights were seen in dams that had been treated with 6 mg/kg-d of PFOS. In addition, eye opening, pinna detachment, and surface righting reflex were all delayed. Female pups from dams exposed to 6 mg/kg-d of PFOS and stressed by immobilization exhibited reduced open-field activity. Male pups did not show any differences in activity, and rotarod performance was unaffected in any group, either by PFOS alone or when PFOS was combined with maternal stress (Figure 3.1).

In a separate study, groups of 4–7 male neonatal (10-day-old) Naval Medical Research Institute (NMRI) mice were administered a single dose of PFOS at 0, 0.75, or 11.3 mg/kg BW by oral gavage (Johansson *et al.* 2008). Following exposure, spontaneous behavior (i.e., locomotion, rearing, and total activity) and habituation were explored in 2- to 4-month-old mice. Motor activity was measured during a 60-minute period, that was divided into three 20-minute sessions. Locomotion, rearing, and total activity were recorded. No effects were observed on body weight. At 2 months old, mice exposed to 0.75 and 11.3 mg/kg BW of PFOS showed significantly decreased rearing, locomotion, and total activity during the first 20 minutes, compared with controls. After 60 minutes, activity was increased in the 11.3 mg/kg BW dose group, when compared with controls. At 4 months, the responses in the 0.75 mg/kg BW dose group were similar to the controls. Thus, a single PFOS treatment at PND 10 affected habituation – even in mice up to 4 months of age in the highest dose group (11.3 mg/kg-d). A LOAEL of 0.75 mg/kg was found from observations of decreased locomotion, rearing, and total activity in 2-month-old mice (Johansson *et al.* 2008).

PFOS: ORAL INGESTION HEALTH EFFECTS TO MAMMALS

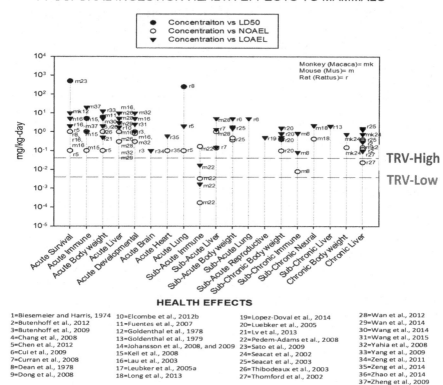

FIGURE 3.1   Oral ingestion health effects of mammalian species.

MAMMALIAN ORAL TOXICITY: CHRONIC TOXICITY – NON-CARCINOGENIC EFFECTS

Only one chronic exposure study had been published for assessment of the chronic effects of PFOS on animals (Thomford 2002, Butenhoff *et al.* 2012). This report represents the long-term component of the Seacat *et al.* (2002) sub-chronic study reported above under the section entitled "Mammalian Oral Toxicity – Sub-Chronic (Rats)". In this work, Sprague Dawley Crl:CD (SD) IGS BR rats (n=40–70) were dosed, using a PFOS-containing diet, for up to 105 weeks. Groups of five rats/sex/dose were sacrificed at weeks 4 and 14 as described above. At termination of the study, it was found that rats treated with PFOS displayed decreased body weight, with increased liver weight, and hepatocellular hypertrophy. A satellite group of animals received 20 ppm of the PFOS-containing diet for 52 weeks, with subsequent feeding of the animals with the control diet until sacrifice at week 106. Animals received dietary levels of 0, 0.5, 2, 5, or 20 ppm PFOS as the potassium salt. Corresponding PFOS doses for male animals were 0, 0.024, 0.098, 0.24, or 0.984 mg/kg-d respectively, and, for females, corresponding doses were 0, 0.029, 0.120, 0.299, or 1.251 mg/kg-d (Seacat *et al.* 2002).

In a similar 2-year study of PFOS effects on rats (Thomford *et al.* 2002), at termination, hepatotoxicity, characterized by centrilobular hypertrophy, centrilobular eosinophilic hepatocytic granules, and centrilobular hepatocytic vacuolation, was noted in rats that received doses of approximately 0.4 or 1.5 mg/kg-d of PFOS in the diet. In addition, rats that received up to approximately 1.5 mg/kg-d of PFOS in the diet for two years had no significant gross or microscopic changes in the heart, and there were no induced morphological changes in the gastrointestinal tract (Thomford 2002). Similarly, no significant hematological effects were reported in this 2-year study in rats that were dosed with up to approximately 1.5 mg/kg-d PFOS in the diet. In addition, rats did not display any significant effects on the gross or microscopic appearance of the femur, sternum, or thigh skeletal muscle (Thomford, 2002).

Five animals/sex in the treated groups were sacrificed during week 53 and liver samples were obtained for laboratory analyses (i.e., mitochondrial activity, hepato-cellular proliferation rate, and determination of palmitoyl-CoA oxidase activity). In addition, liver weights were recorded. Serum specimens were collected at weeks 27 and 53 from 10 rats/sex/dose group to determine potential clinical effects associated with systemic toxicity. Liver samples were also obtained both during and at the end of the study to quantify levels of PFOS. Data on chronic effects were not reported for the recovery group (Butenhoff *et al.* 2012).

The serological levels of PFOS were measured at weeks 4, 14, and 105. With the exception of the 0.5 ppm exposure group, levels of PFOS in the sera of male rats decreased between week 14 and week 105 by almost 50%. In the 0.5 ppm group, a much greater decrease in serological levels of PFOS was observed over this period. A serum measurement was available at 53 weeks for the high-dose-exposed males and was comparable to the value that was found at 14 weeks. In female rats, serum levels remained relatively constant at weeks 14 and 105. In both males and females, the concentrations of PFOS in the liver were lower at 105 weeks than at 14 weeks (Butenhoff *et al.* 2012).

The clinical serum observations for ALT activity at 53 weeks were consistent with those determined at 14 weeks and showed significant ($P < 0.05$) increases for the high-dose PFOS-treated males but not the corresponding female animals. At week 27, ALT activity was increased in high-dose males, although the difference was not statistically significance. For male animals at 53 weeks in the 0, 0.5, 2, 5, and 20 ppm groups, mean ± SD ALT values were 54 ± 66, 62 ± 52, 40 ± 7.5, 44 ± 8.3, and 83 ± 84 IU/L, respectively. The data suggest that some animals might display greater sensitivity to liver damage following PFOS exposure than others. AST activities remained unaltered in both male and female animals, regardless of PFOS treatment. Serum blood urea nitrogen (BUN) was significantly ($P \leq 0.05$) increased in the 20 ppm-dosed male and female animals for weeks 14, 27, and 53, and in the 5 ppm-dosed male and female animals at weeks 27 and 53 (Butenhoff *et al.* 2012).

Non-neoplastic lesions in the liver of both male and female animals differentially included centrilobular hypertrophy, the presence of eosinophilic granules, vacuolation, and single cell necrosis. At sacrifice, males dosed at 2 ppm exhibited a significant ($P < 0.05$) increase in hepatocellular centrilobular hypertrophy. In the male and female animals at 5 and 20 ppm, there were significant ($P < 0.05$) increases in centrilobular hypertrophy, centrilobular eosinophilic hepatocytic granules (females only), and centrilobular hepatocytic vacuolation (males only). At the high PFOS dose, there was a significant increase in the number of animals with single-cell hepatic necrosis in both male and female rats at 53 weeks. Necrosis in the animals during recovery was comparable to that seen in the control animals.

No effects were observed on hepatic palmitoyl-CoA oxidase activity or increases in proliferative cell nuclear antigen (PCNA) at weeks 4 and 14 or bromodeoxyuridine at week 53. PFOS was detected in the liver and serum samples of the treated animals, with trace amounts being identified in the control animals. The LOAEL at termination for male rats was 2 ppm (0.098 mg/kg-d) and was 5 ppm (0.299 mg/kg-d), based on the liver histopathology. The NOAEL for the males was 0.5 ppm (0.024 mg/kg-d) and 2 ppm (0.120 mg/kg-d) for females. Additional details from the study, with regard to carcinogenicity, are described below.

A summary of the chronic oral toxicological effects of PFOS in mammalian species is shown in Table 3.3.

TABLE 3.3
Summary of Chronic Oral Toxicity for PFOS in Mammalian Species

Test Organism	Test Duration	Test Results		Effects Observed at the LOAEL	Reference
		NOAEL (mg/ kg-d)	LOAEL (mg/ kg-d)		
Monkey	180 days	0.15	0.75	Based on decreased body weight, increased liver organ weight, and decreased levels of cholesterol.	Seacat *et al.* 2002
Rat	98 days	Females 0.40 Males 0.34	Females 1.56 Males 1.33	Based on increased liver organ weights, decreased levels of cholesterol (males), increased activity of ALT (males), increased BUN (males and females), increased liver hypertrophy, and hepatic centrilobular vacuolization.	Seacat *et al.* 2003
Rat	104 weeks	Females 0.40 Males 0.34	Females 1.56 Males 1.33	Derived from examination of liver histopathology specimens Derived from examination of liver histopathology specimens Observations included cystic degeneration, centrilobular vacuolation (females), centrilobular eosinophilic granules (males), increased hepatic necrosis, and centrilobular vacuolation at high doses. Carcinogenicity of exposure to PFOS.	Thomford *et al.* 2002 Butenhoff *et al.* 2012

Key: ALT, BUN

MAMMALIAN INHALATION TOXICITY

Mammalian Inhalation Toxicity: Acute

Data that were derived from an LC_{50} study of the median acute lethal concentration for 50 percent of the exposed animals in rats suggests that PFOS absorption can take place by an inhalational exposure route (Rusch *et al.* 1979). Since the tools and approaches for quantifying PFOS in animals were limited at the time the study was published, the published report did not include pharmacokinetic data.

The acute inhalation toxicity of PFOS was conducted in SD rats (5/sex/dose) exposed to PFOS dust in air at concentrations of 0, 1.89, 2.86, 4.88, 6.49, 7.05, 13.9, 24.09, or 45.97 mg/L for 1 hour. Rats were observed for abnormal outcomes pre-exposure, at 15-minute intervals during the exposure, at removal from the chamber, hourly for 4 hours post-exposure and then finally each day for up to 14 days post-exposure (Rusch *et al.* 1979). It should be noted that the 45.97 mg/L group was not used in determining the LC_{50} as this treatment group of the study was terminated on day 2 due to high mortality; the 13.9 mg/L group was also not part of the calculation as this group was terminated early due to mechanical problems. All rats in the 24.09 mg/L group died by day 6 post-exposure, and mortality observed for the other groups after 14 days was 0, 10, 20, 80, and 80 per cent in the 1.89, 2.86, 4.88, 6.49, and 7.05 mg/L dosed groups, respectively. Observed clinical signs from respiratory exposure included emaciation, red material around the nose or nasal discharges, dry rales, and breathing disturbances, or generally poor conditioning. Necropsy indicated discoloration of the liver and lungs. Based on the observations described above, the acute rat inhalation LC_{50} value for PFOS was 5.2 mg/L (ppm) (Rusch *et al.* 1979). In addition, unpublished data (OECD, 2002) indicated that exposure of male and female SD rats to concentrations of PFOS dusts between 1,890 and 45,970 mg/m^3 for 1 hour induced dry rales and other breathing disturbances, and distension of the small intestine when exposed to lethal concentrations of PFOS. An LC_{50} of 5,200 mg/m^3 was calculated for PFOS from these data (OECD 2002, ATSDR 2015).

MAMMALIAN SUB-ACUTE INHALATION

No sub-acute inhalation PFOS toxicity data were identified in the literature search.

Mammalian Sub-Chronic Inhalation

No studies on sub-chronic inhalation exposure with PFOS were available.

Mammalian Chronic inhalation

No chronic inhalation toxicity data from PFOS studies were available.

MAMMALIAN DERMAL/OCULAR TOXICITY

The acute dermal and ocular irritation test with PFOS was performed in an albino rabbit model (Biesemeier and Harris 1974, summarized in OECD 2002). This study represents the only mammalian test of dermal or ocular irritancy conducted with

PFOS, in which six rabbits were treated by placing 0.5 g of the test article on intact or abraded backs of the animals, which were then covered. Following exposure, erythema and edema were scored after 24 and 72 hours. The primary dermal irritation score from this study was zero, which indicated no skin irritation potentials or edema from PFOS exposure in this model. Unfortunately, technical details were lacking in the report with regard to the carrier vehicle of PFOS used, the gender of the animals, or of the technical guidelines that were followed (Biesemeier and Harris 1974).

In the ocular irritation component of the study, six albino test rabbits (New Zealand White) were treated with 0.1 g of the test article in one eye, and the remaining eye was left untouched as an untreated control. Any reaction to PFOS was recorded at 1, 24, 48, and 72 hours after treatment. It was noted that scale criteria were unavailable or referenced for this study. Scores were maximal at 1 hour and 24 hours post-treatment, which then decreased over the duration of the study, and PFOS was found not to be an eye irritant. Additionally, raw data were unavailable in the OECD (2002) report.

In acute oral and contact toxicity evaluations of PFOS exposure, the honey bee (*Apis mellifera*) was studied (Wilkins 2001a and 2001b). Honey bees were selected because of their critical importance as pollinators of several diverse agricultural crops and are commonly used as a model species in determining pesticide toxicity (Beach *et al.* 2006). The acute oral toxicity evaluation of PFOS was conducted using honey bees that were exposed to 0.21, 0.45, 0.99, 2.17, or 4.78 µg PFOS/bee in an aqueous sucrose carrier solution for 4 hours. No mortality or sub-lethal effects were noted following treatment with 0.21 µg PFOS/bee. The 72-h oral LD_{50} was calculated at 0.4 µg PFOS/bee with a slope of 4.8 (Wilkins *et al.* 2001b). If the LD_{50} is converted to dose per kg of food, the value becomes 2.0 mg PFOS/kg sugar solution. Based on the 72-h oral LD_{50}, PFOS was calculated as "highly toxic" by the International Commission for Bee Botany (ICBB, Wilkins 2001a).

Furthermore, in the acute contact study, treatments were administered by applying doses of 1.93, 4.24, 9.3, 20.5, or 45 µg PFOS in acetone per bee to the thorax of bees. Based on mortality data, the estimated LD_{50} at 96 hours was 4.8 µg PFOS/bee, and the NOAEL was 1.93 µg PFOS/bee. The slope of the dose-response line was estimated to be 4.0 and the 96-h LD_{50} was determined to be 4.8 µg PFOS/bee. Based on the 96-h contact LD_{50}, PFOS was classified as moderately toxic by the ICBB (Wilkins 2001b). No sub-chronic or chronic dermal toxicity data were available.

MAMMALIAN TOXICITY – OTHER: GENOTOXICITY/MUTAGENICITY

In general terms, chromosomal aberrations and genome instability are among the initial cellular responses that are provoked by environmental toxicants that are also carcinogenic – collectively these processes ultimately provoke an increased risk of initiating carcinogenesis and cancer progression. However, many studies have explored the potential genotoxicity of PFOS by complementary techniques, which collectively indicate that it does not directly induce DNA damage (Oda *et al.* 2007, Kawamoto *et al.* 2010). For example, when using the Umu Chromotest, it was previously shown that PFOS (as well as PFOA and fluorotelomer alcohols) did not significantly increase β-galactosidase activity over the concentration range of 0–1,000 μM in the presence or absence of the S9 liver fraction metabolic activation mix (Oda *et al.* 2007).

In another related study, Kawamoto *et al.* (2010) also confirmed that PFOS does not directly induce DNA damage. In this particular study, the potential genotoxicity of PFOS and other related compounds was assayed by an *in vivo* comet assay developed for *Paramecium caudatum*, in which PFOS failed to induce DNA damage under the conditions tested (Kawamoto et al. 2010). Additionally, in work reported by the USEPA (2014) and the European Food Safety Authority (EFSA 2008), genotoxicity assays of PFOS showed negative results from the traditional Ames mutation assay, which showed that PFOS is a non-mutagenic agent, based on interpretation of the negative results obtained in a reverse mutation test in *Salmonella typhimurium* (EFSA 2008) and, with regard to *in vitro* assays of chromosomal aberrations, unscheduled DNA synthesis assays, and *in vivo* mouse micronucleus assays (USEPA 2014). For this reason, observed findings of a potential association between PFOS exposure and cancer are postulated to be caused by non-genotoxic mechanisms (Wang *et al.* 2015b).

However, the above experimental evidence is balanced by studies showing the ability of PFOS to intercalate into DNA and cause DNA base damage in a cell-free system (Lu *et al.* 2012). Furthermore, PFOS is capable of inducing strand breaks and fragmentation of genomic DNA in green mussels and earthworms (Liu *et al.* 2014, Xu *et al.* 2013). Additionally, the potential mutagenic effects of PFOS were illustrated in a novel *gpt* delta (guanine phosphoribosyltransferase gene) transgenic mouse mutation system, that revealed a role for hydrogen peroxide (H_2O_2) in PFOS-induced genotoxicity (Wang *et al.* 2015). Furthermore, mutations at the *redBA/gam* loci were determined by the Spi⁻ assay both *in vitro* and *in vivo*.

DNA damage was shown by the phosphorylated histone H2AX (γ-H2AX) and murine bone marrow micronucleus (MN) assays, which showed that PFOS induced concentration-dependent increases in γ-H2AX foci and in mutation frequencies at *redBA/gam* loci in transgenic mouse embryonic fibroblast cells, which were confirmed by the formation of micronuclei in the bone marrow, and induced mutations in the livers of *gpt* delta transgenic mice. Moreover, the secretion of H_2O_2 was closely associated with abnormal peroxisomal β-oxidation that was caused by PFOS – observations that the authors claimed provided new mechanistic information with regard to the genotoxic effects of PFOS (Wang *et al.* 2015b).

Furthermore, in one particular study in which PFOS was administered orally to rats at doses of 0.6, 1.25, or 2.5 mg/kg, investigators noted increased micronucleus frequency and PFOS-induced DNA damage in the peripheral blood (Eke and Celik 2019). A linear dose-response relationship was seen between PFOS concentration and the frequency of micronucleated cells, with all doses of PFOS inducing DNA damage. PFOS-treated cells showed a statistically significant and dose-dependent increase in the genetic damage index (arbitrary units, AU) and mean damaged cell percentage (DCP) values as compared with the negative control ($P < 0.001$; Eke and Celik 2019).

In addition to limited and relevant genotoxicity data in the context of wildlife species, these are some additional data that suggest genotoxic effects on aquatic organisms. For example, the toxicological effects of PFOS in the common carp (*Cyprinus carpio*) were evaluated by determining the responses of five key biomarkers, including DNA single-strand breaks (Kim *et al.* 2010). This study found that a significant increase in DNA damage occurred on exposure to PFOS, while other markers were not significantly affected.

MAMMALIAN TOXICITY – OTHER SYSTEMS AND EFFECTS

Immunotoxicity and PFOS

Several studies have reported PFOS-associated immunotoxicity in animal studies of innate and adaptive immune systems, although the available data indicates inconsistent or controversial results from rat and mouse studies (DeWitt *et al.* 2012, Corsini *et al.* 2012). These studies have indicated that mice are markedly more sensitive than rats to PFOS and other perfluoroalkyl compounds. In mouse model studies, it was shown that PFOS affects antibody synthesis at levels found in the general human population (Peden-Adams *et al.* 2008). It was shown that serological levels of PFOS that were associated with the LOAEL for blunted humoral immunity were 91.5 ng/g (at an administered PFOS dose of 0.05 mg/kg over the course of 28 days) for males and 666 ng/g (at an administered PFOS dose of 0.5 mg/kg over the course of 28 days) for females (Peden-Adams *et al.* 2008). In addition, a 21-day oral gavage exposure to PFOS at doses of 5 or 25 μg/kg PFOS/kg-d (resulting in 189 or 670 ng/mL plasma PFOS, respectively) suppressed immunity to influenza A virus (H1N1 serotype) infection in mice, and resulted in marked increases in emaciation and mortality (Guruge *et al.* 2009). The pesticide sulfluramid, which is rapidly metabolized to PFOS, can also target T-dependent IgM antibody production in mice at exposure levels that are ten-fold lower than overt toxic doses (Peden-Adams *et al.* 2007).

In a study of B6C3F1 mouse pups born from the pairing of female C57BL/6N with male C4H/HeJ mice, the dams were gavaged with 0.1, 1, or 5 mg/kg-d PFOS at GD 1–17, and immunological effects determined (Keil *et al.* 2008). Natural killer (NK) cell activity was suppressed in 8-week-old male offspring at the 1 and 5 mg/kg-d doses and in female offspring at the 5 mg/kg-d dose (Keil *et al.* 2008). Specific levels of IgM were also suppressed at the 5 mg/kg-d dose level, which was sufficient to cause hepatomegaly at four weeks in male mice, but not in females. Serum PFOS concentrations were not reported in this study (Keil *et al.* 2008). However, others have found no effect following dietary exposure of male B6C3F1 mice to 7 mg/kg-d PFOS for 28 days (Qazi *et al.* 2010). Whereas this dosing schedule gave a serum concentration of 11,600 ng/mL PFOS (representing more than 10 times the average concentration typically found in occupationally exposed humans), PFOS exposure did not alter IgG or IgM synthesis (Qazi *et al.* 2010). This study also found no effect of PFOS on the total number of circulating leukocytes or the number and phenotypic distribution of thymic or splenic cells.

Furthermore, the immune systems of non-mammalian wildlife species, including birds, turtles, fish, and lizards, can be modulated at current environmental levels of PFCs (Peden-Adams *et al.* 2009). Roland *et al.* (2014) studied the impact of the chronic exposure (i.e., 28-days exposure) of PFOS (at 0, 1, or 10 µg/L) in the European eel (*Anguilla anguilla*), and explored any subsequent effect on the functional activity and immunophenotype of peripheral blood mononuclear cells (PBMC). The authors found that 17 proteins were differentially expressed on exposure of eels to 1 or 10 µg/L PFOS. The functional classes mainly represented were cytoskeleton and carbohydrate metabolism processes (Roland *et al.* 2014).

In terms of cell-mediated innate immunity, there is only moderate confidence that PFOS exposure alters NK cell activity in animals. NK cells are important for resistance against viruses and tumor cells. Successful defense by NK cells involves killing of target cells through release of cytolytic granules or induction of apoptosis (Dietert *et al.* 2010). Assays for NK cell activity are included in many immunotoxicity testing guidelines as a measure of immune function, because they are considered to be good predictors for overall immunotoxicity (Luster *et al.* 1992, USEPA 1998, ICH 2005, WHO 2012).

There is consistent evidence that PFOS exposure results in suppression of NK cell activity in mice at doses from 0.833 to 40 mg/kg-d PFOS (Keil *et al.* 2008, Dong *et al.* 2009b, Zheng *et al.* 2009, Vetvicka and Vetvickova 2013). However, at lower doses (0.0166 to 0.166 mg/kg-d), the results are mixed, including no effects of PFOS (Keil *et al.* 2008), no effect on female mice Peden-Adams *et al.* 2008) or increased NK cell activity (in male mice; Peden-Adams *et al.* 2008, Dong *et al.* 2009b). The LOAEL after developmental exposure (1 mg/kg-d) (Keil *et al.* 2008) is similar to LOAEL values reported following exposure of juveniles or adults. Furthermore, the suppression of NK activity at doses above 0.833 mg/kg-d PFOS was considered a consistent pattern of findings and not downgraded for inconsistent effects at lower doses. Reduced NK cell activity in the lower dose range (0.833 to 5 mg/kg-d PFOS) appears to occur without changes in body weight, spleen or thymus cellularity, or other signs of overt toxicity.

The conclusions are restricted to the evidence of a PFOS-associated suppression in NK cell activity in mice. However, when tested in marine mammals, others have shown an increasing trend in *ex vivo* PFOS-stimulated bottlenose dolphin (*Tursiops truncates*) B-cell proliferation, while ex vivo PFOS-stimulated NK-cell activity and lymphocyte proliferation was unaltered at all doses tested (0.01, 0.05, 0.1, 0.5, 1.0, and 5.0 µg/ml) as compared controls (Wirth et al. 2014). This same group also confirmed that B6C3F1 female mouse splenocytes showed no altered B-cell proliferation; however, T-cell proliferation was decreased at all PFOS doses and NK-cell activity was decreased at 0.01, 0.05, 0.1, 0.5, and 1 µg/ml and increased at 5 µg ml–1 PFOS. Suggestive of a U-shaped dose-response affect in this model (Wirth *et al.* 2014)

Additional data would be necessary to further characterize the shape of the dose-response curve at lower doses. Although NK-cell activity was suppressed at doses above 0.833 mg/kg-d PFOS, confidence was not increased for dose response as the evidence for dose response was unclear. Several studies showed the same or lower magnitude of effect at higher doses (e.g., male mice showed 42% reduction in NK activity at 1 mg/kg dose, but only a 28% reduction at 5 mg/kg in Keil *et al.* (2008)).

Wildlife studies provided mixed evidence of a PFOS effect. Higher concentrations of PFOS were found in sea otters (*Enhydra lutris*), with clinical signs of disease compared to healthy animals (Kannan *et al.* 2006). In contrast, no difference in PFOS levels was seen between little brown bats (*Myotis lucifugus*) in the population, with respect to white-nose syndrome, as compared with the healthy reference population (Kannan *et al.* 2010). Further, in a wildlife PFOS dose-dependency study, no association was found between high serum levels of PFOS (mean 1,420 ng/mL; range 317–6,257 ng/mL) and NK cell activity (n = 12 per dose group) in bottlenose dolphins (Fair *et al.* 2013). It should be noted that the animals used for this study were from captive and thus carefully managed populations. In addition, serum PFOS levels of the dolphins used as sources of NK-cells were not tested and background levels of PFOS may have decreased the sensitivity of the assay and its ability to detect a possible effect of PFOS. It is possible for wild-caught dolphins to display very high levels of serum PFOS, an observation that is due in part to their diet (i.e., from 317 to 6,257 ng/mL), as previously reported (Fair *et al.* 2013).

Others have previously shown that PFOS provokes inflammation in appropriate animal models (Dong *et al.* 2012). In a standard 60-day oral exposure study, adult male C57BL/6 mice were dose-dependently exposed to PFOS by oral gavage at 0, 0.0083, 0.0167, 0.0833, 0.4167, 0.8333, or 2.0833 mg/kg-d (Dong *et al.* 2012). Over 60 days, this regimen yielded a targeted Total Administered Dose (TAD) of 0, 0.5, 1, 5, 25, 50, or 125 mg PFOS/kg, respectively. This study identified a previously under-appreciated mode of action of PFOS, wherein it markedly increased the frequency of peritoneal macrophages (CD11b cells) at PFOS concentrations of 1 mg/kg TAD, doing so a dose-dependent manner. *Ex vivo* interleukin IL-1β production by peritoneal macrophages was elevated at concentrations of 5 mg/kg TAD. Moreover, PFOS exposure markedly enhanced the *ex vivo* secretion of tumor necrosis factor TNF-α, IL-1β, and IL-6 by peritoneal and splenic macrophages when stimulated by lipopolysaccharides (LPS, or endotoxins), both *in vitro* or *in vivo*. The serum levels of these inflammatory cytokines, that were observed in response to *in vivo* stimulation by LPS, were elevated substantially by exposure to PFOS. In addition, PFOS exposure elevated the expression of the genes encoding TNF-α, IL-1β, IL-6, and the proto-oncogene, c-myc, in the spleen. These data suggest that exposure to PFOS modulates the inflammatory response. Clearly, further research is needed to determine the mechanism of action (Dong *et al.* 2012).

In sum, animal studies indicate, on balance, that higher serum levels of PFOS are associated with suppressed antibody responsiveness. There is marked confidence that PFOS exposure is associated with dampened antibody responses in animals based on experimental studies carried out on mice, where consistent suppression of the primary antibody response was seen. Reduced antibody production is an indication of decreased immune function, immunocompetence, or immunosuppression that may indicate a greater risk of disease susceptibility. Moreover, consistent evidence indicates that PFOS exposure dampens primary antibody responsiveness, as determined by antigen-specific IgM antibody production to a single challenge with T-cell-specific antigens (e.g., sheep red blood cells or SRBC) in both male and female mice (Keil *et al.* 2008, Peden-Adams *et al.* 2008, Dong *et al.* 2009, Zheng *et al.* 2009, Qazi *et al.* 2010, Dong *et al.* 2011, Vetvicka and Vetvickova 2013).

This observation was supported by a study carried out in chickens (Peden-Adams *et al.* 2009) at oral doses ranging from 0.00166 to 40 mg/kg-d. Antibody suppression in the lower dose range (i.e., 0.00166 to 5 mg/kg-d PFOS) occurs without evidence of altered body weight, spleen or thymus cellularity, or other signs of overt toxicity. Overall, PFOS-induced immunological alterations in adult mice were characterized by thymic and splenic atrophy, modulated thymocyte and splenocyte phenotypes, and immunocompromised responses to T-cell-dependent stimulatory antigens. For PFOS, several studies derived LOAELs of 0.02–0.8 mg/kg-d, whereas one study identified a LOAEL of 0.00166 mg/kg-d for suppressed responsiveness to a T-cell-dependent antigen.

SUMMARY OF AVIAN TOXICOLOGY

ACUTE TOXICITY

The acute oral toxicity of dietary PFOS was examined in 10-day-old (juvenile) mallard (*Anas platyrhynchos*) and northern bobwhite (*Colinus virginianus*) in a 5-day study (Newsted *et al.* 2006). The birds were then observed for three days while being fed uncontaminated food, following which half of the birds were sacrificed at day 8 of the study. The remaining birds were all maintained on the PFOS-free basal ration for an additional two weeks and then sacrificed on day 22 of the study. Birds were evaluated for growth, rate of feed consumption, behavior, physical injury, gross abnormalities, and mortality. In addition, liver weights and PFOS levels in the sera and liver were determined. Target nominal diet concentrations for mallard were 0 (control), 8.8, 17.6, 35.1, 70.3, 141, 281, 562, or 1,125 mg PFOS/kg feed. For northern bobwhite, target nominal concentrations were 0, 17.6, 35.1, 70.3, 141, 281, 562, or 1,125 mg PFOS/kg feed (Gallagher *et al.* 2003a, 2003b, Newsted *et al.* 2006). Thus, for mallard, the estimated average daily intake (ADI) values associated with nominal exposures to PFOS were 2.71, 5.74, 12.0, 61.3, 74.2, 149, and 228 mg/kg-d, respectively, and, for bobwhite, the estimated ADI values associated with the nominal exposure were 4.83, 8.52, 23.8, 44.7, 76.4, 193, and 225 mg/kg-d, respectively (Newsted *et al.* 2006).

In the acute dietary study, treatment-related mortalities were observed in juvenile mallard from the 281, 562, and 1,125 mg PFOS/kg feed treatments, starting at days 7, 5, and 4 of dosing, respectively. No mortality signs were observed at 141 mg/kg. Overt signs of toxicity in mallard from the 281, 562, and 1,125 mg PFOS/kg feed treatment were observed, starting at days 4, 3, and 2 of dosing, respectively. Signs of toxicity included ruffled appearance, reduced reaction to stimuli, lethargy, loss of coordination, prostrate posture, convulsions, and lower limb weakness. After day 8, no further mortality was observed in the study. Mortality and signs of toxicity were not observed at doses of 141 mg PFOS/kg feed or lower at any phase in the study (Newsted *et al.* 2006).

Based on the ADI of PFOS calculated over the 5-day exposure period, the LD_{50} for juvenile mallard was determined to be 150 mg PFOS/kg BW-day, equivalent to a total cumulative dose of 750 mg PFOS/kg BW, calculated over a 5-day period. For juvenile bobwhite, the LD_{50} based on the ADI was 61 mg PFOS/kg BW-day, equivalent to a total cumulative dose of 305 mg PFOS/kg BW. Reductions in weight gain and body weight were observed in bobwhite from the 141 mg PFOS/kg treatment, but these measures returned to control levels by Day 22. The no-mortality dietary treatments were 70.3 and 141 mg PFOS/kg feed for bobwhite and mallard, respectively. Both mallard and bobwhite accumulated PFOS in blood serum and liver in a dose-dependent manner. The half-lives of PFOS in mallard blood serum and liver were estimated to be 6.86 and 17.5 days, respectively, whereas, in bobwhite, the half-life of PFOS in liver was estimated to be 12.8 days, while the half-life of PFOS in bobwhite blood serum could not be estimated. Concentrations of PFOS in livers of juvenile mallard and bobwhite associated with the observed mortality were at least 50-fold greater than the single maximum PFOS concentration that has been measured in livers of avian wildlife. Although no treatment-related effects on body weight were observed in bobwhite quail that had been exposed to concentrations less than or equal to 70.3 mg PFOS/kg feed, bobwhite appeared to be more sensitive than mallard to the acute toxicological effects of PFOS.

The pharmacokinetics and potential for acute lethality of PFOS to juvenile mallard and northern bobwhite was explored by Newsted *et al.* (2006). Ten-day-old mallard and northern bobwhite were fed PFOS in the diet for 5 days. Birds were then observed for an additional 3 days, whilst being fed uncontaminated feed. Half of the birds were sacrificed on day 8 of the study.

The kinetics of elimination and tissue disposition of PFOA and PFOS were determined in 6-week old male chickens that had been exposed by subcutaneous implantation (Yoo *et al.* 2009). Chickens were exposed to either PFOA (at 0.1 or 0.5 mg/mL) or PFOS (at 2.0×10^{-2} or 0.1 mg/mL), or a saline vehicle control. Chickens were exposed to both doses of each test article for 4 weeks and then allowed to depurate for 4 weeks.

The concentrations of total cholesterol and phospholipids were lower in chickens that had been exposed to PFOS, although other measures of clinical biochemistry, body index, and histology were not significantly different from the controls. The elimination rate constant for PFOA was 0.15/day and for PFOS it was 0.023/day. The highest concentrations of PFOA and PFOS were found in the kidneys and liver, respectively. The half-life for depuration of PFOA was 4.6 days and for PFOS it was 125 days (Yoo *et al.* 2009).

Other researchers have reported the bioaccumulation in nature of PFOS in the liver and blood of the great tit (*Parus major*), which is a small songbird found throughout western Europe (Dauwe *et al.* 2007). This study found that PFOS concentrations ranged from 553 ng/g to 11,359 ng/g in the liver and ranged from 24 to 1,625 ng/mL in the blood. At the time of this study, these levels were among the highest ever reported in free-living animals and exceeded the previously published hepatic benchmark concentrations that were protective of avian species in almost all birds (Beach *et al.* 2006). A summary of the acute toxicity of PFOS in avian species is described in Table 3.4.

TABLE 3.4
Summary of Acute Oral Toxicity for PFOS in Avian Species

| | | | | Test Results | | |
	Test Duration	LD$_{50}$ (mg/kg)	NOAEL (mg/kg-d)	LOAEL (mg/kg-d)	Effects Observed at the LOAEL	Reference
Test Organism						
Northern bobwhite (*Colinus virginianus*)	8-day dietary study	61	24	NA	Based on post-exposure lethality in juvenile birds	Newsted *et al.* 2005
Mallard (*Anas platyrhynchos*)	8-day dietary study	150	61	NA	Based on post-exposure lethality in juvenile birds	
Northern bobwhite	8-day dietary study	NA	24 (ADI)	48 (ADI)	Based on reductions in whole body weight in juvenile birds	Newsted *et al.* 2005 and 2006
Mallard (*Anas platyrhynchos*)	8-day dietary study	NA	12 (ADI)	20 (ADI)	weight in juvenile birds Based on reductions in whole body weight in juvenile birds	

Key: NA = not applicable ADI

SUB-CHRONIC TOXICITY IN AVIAN SPECIES

No sub-chronic studies on birds were identified when searching the available literature (Figure 3.2).

CHRONIC TOXICITY IN AVIAN SPECIES

Chronic effects of PFOS on adult mallard ducks and northern bobwhite were studied
following exposure of the birds to PFOS in the diet (Newsted *et al.* 2007). Adult mallard ducks and northern bobwhite (male and female) were exposed to 0, 10, 20, 50,
or 150 mg of PFOS/kg of diet for 21 weeks. The measured mean feed concentrations
of PFOS were 10.2, 20.8, 50.9, and 161 mg PFOS/kg feed for the nominal dietary
concentrations of 10, 20, 50, and 150 mg PFOS/kg feed, respectively (Newsted *et al.*
2007). The average daily intake of PFOS mg/kg was 0.77, 2.64, or 7.32 mg/kg BW
for bobwhite and 1.48, 6.36, or 20.9 mg/kg BW for mallard at the respective nominal
dietary PFOS concentrations of 10, 50, and 150 mg PFOS/kg feed, respectively.

PFOS: ORAL INGESTION HEALTH EFFECTS IN AVIAN SPECIES

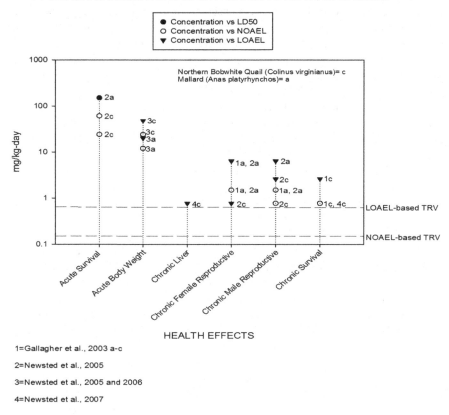

FIGURE 3.2   PFOS: Health effects on avian species.

Due to the observed mortality, the study of birds exposed to 50 or 150 mg PFOS/ kg feed was terminated by week 7, and birds in the 150 mg/kg feed dose group had a dose-reduced exposure to PFOS of 20 mg/kg feed. There was no mortality observed in either mallard or bobwhite during the study in the 10 mg/kg feed group. No treatment-related clinical effects were seen in mallard; however, in bobwhite, clinical signs, lethargy, and reduced reaction to external stimuli, and a ruffled appearance were seen starting at week 5 (Newsted *et al.* 2007). No apparent treatment-related effects on body weight were seen for bobwhite or mallard exposed to 10 mg/kg feed but a significant reduction in body weight in the higher dose groups was seen in both species of bird.

There was a statistically significant increase in liver weights of female bobwhite in the 10 mg/kg feed group but treatment-related effects were not observed in the livers of mallard at this dose. No livers were collected from the higher dose group birds. Gross or histological changes were not observed in female bobwhite or mallard; however, male bobwhite and mallard in the 10 mg PFOS/kg feed group showed an incidence of smaller (by length) testes as compared with controls. However, spermatogenesis was unaffected and there was no effect on the rate of egg fertilization. No treatment-related effects were observed in histopathological evaluation of the liver, kidney, brain, proventriculus, and bursa of Fabricius from mallard and bobwhite of the 10 mg/kg feed group that were exposed for 21 weeks. The NOAEL for adult male bobwhite and male mallard was determined to be 10 mg/kg feed for the 21-week feeding duration. For adult male birds, a LOAEL was not determined for either species of bird (Newsted *et al.* 2007).

Based on reproduction studies (Gallagher *et al.* 2003a, 2003b), the LOAEL was derived only for the female mallard, which was 50 mg PFOS/kg in the diet (6.36 mg PFOS/kg BW-d), whereas the NOAEL was 10 mg PFOS/kg in the diet for both the male and female mallard (1.49 PFOS/kg BW-d).

For adult female bobwhite, the LOAEL was calculated to be 10 mg PFOS/kg wet weight (WW) diet (0.77 PFOS/kg BW-d), which was based on a decreased survival of 14-day-old bobwhite offspring (i.e., survivors as a percentage of eggs set by quail that had been fed PFOS in the diet) – in this study, a NOAEL was not derived. For adult male bobwhite, the LOAEL, based on effects on body weight and mortality, was 50 mg PFOS/kg diet (2.64 mg PFOS/kg BW-d), whereas the NOAEL was calculated as 10 mg PFOS/kg in the diet (0.77 mg PFOS/kg BW-d). Based on slight effects observed in birds that were fed 10 mg PFOS/kg feed in the definitive study, and when considered in conjunction with the pilot study results (Gallagher et al. 2003c), the bird populations that were exposed to dietary concentrations of PFOS ≤ 6.2 mg PFOS/kg feed (a no-effect level) would not be expected to experience adverse effects on the survival of the adults of the offspring, or reproductive performance of the adults, or the survival, growth, or development of the offspring (Newsted *et al.* 2007).

Furthermore, Beach *et al.* (2006) stated that, for avian species, dietary, ADI, and egg yolk-based benchmarks were determined as 0.28 mg PFOS/kg diet, 0.021 mg PFOS/kg BW-d, and 1.7 μgPFOS/mL yolk, respectively. Benchmarks for serum and liver for the protection of avian species were 1.0 μg PFOS/mL and 0.6 μg PFOS/g WW, respectively. However, no-effect levels in laboratory studies suggest that actual population-level effects would not be expected to occur until a concentration of 6.0 mg PFOS/kg in the diet, 5.0 μg PFOS/g WW in the liver, or 9.0 μg PFOS/mL in the sera was exceeded – levels indicative of the conservative nature of the benchmarks.

A summary of the chronic oral toxicological effects of PFOS on avian species is presented in Table 3.5.

REPRODUCTIVE STUDIES IN AVIAN SPECIES

Developmental toxicity studies in eggs of White Leghorn chickens have revealed immunotoxic and developmental effects following *in ovo* delivery of PFOS (Peden-Adams *et al.* 2009). Eggs were injected with PFOS at concentrations of 1, 2.5, or 5 mg/kg egg weight in a carrier vehicle of sunflower oil/10% dimethyl sulfoxide (DMSO) or the carrier vehicle alone, following which chicks were allowed to develop to post-hatch day (PHD) 14. PFOS exposure did not affect the hatch rate in this model. Following *in ovo* exposure to PFOS, chicks displayed increased spleen mass at all PFOS concentrations, and increased liver mass at 2.5 or 5 mg/kg egg weight

TABLE 3.5

Summary of Chronic Oral Toxicity for PFOS in Avian Species

		Test Results			
Test Organism	Test Duration	NOAEL (mg/kg-d)	LOAEL (mg/kg-d)	Effects Observed at the LOAEL	Reference
Northern bobwhite (*Colinus virginianus*)	21 weeks	Adult males 0.772 Adult females 0.770	Adult males NA Adult females NA	Based on a decrease in the survival of offspring after 14 days and a statistically significant increase in adult female liver weights.	Newsted *et al.* 2007
Mallard (*Anas platyrhynchos*)	21 weeks	Adult males 1.49 Adult females 1.49	Adult males NA Adult females NA		
Northern bobwhite	21 weeks	Adult males and females 0.77	Adults males and females 2.64	Decrease in 14-day survival	Gallagher *et al.* 2003a, b
Mallard	21 weeks	Adult males 1.49 Adults females 1.49	Adults males NA Adult females 6.36	In the reproduction study, no treatment-related mortalities or effects on body weight, weight gain, histopathology, or reproductive measures were seen.	
Northern bobwhite	21 weeks reproduction study	Adult males 0.77 Adults females NA	Adults males 2.60 Adult females 0.77	Based on reproductive studies of up to 21 weeks conducted with adult quail exposed to PFOS in the diet	Newsted *et al.* 2005
Mallard	21 weeks reproduction study	Adult males and females 1.50	Adults males and females 6.40	Based on reproductive studies of up to 21 weeks conducted with adult mallard exposed to PFOS in the diet	

Key: NA = not applicable

and body length (as defined by the crown-rump length) at the 5 mg/kg treatment. All treatment concentrations also resulted in shortened right wings compared to the control. Moreover, titers of sheep red blood cell (SRBC)-specific immunoglobulin (IgM and IgY combined) decreased at all PFOS treatment levels, and plasma lysozyme activity increased in response to all treatment levels (Peden-Adams *et al.* 2009). Crucially, serum concentrations, where significant immunotoxic, morphological, and neurological effects were observed at the lowest dose (1 mg/kg egg weight), averaged 154 ng PFOS/g serum.

It was concluded that these concentrations fall within environmental ranges that have been reported in blood samples from wild-caught avian species. Thus, the observations described above confirm that the environmental serum PFOS concentrations found in hatchlings agree closely with the serum levels found in hatchlings treated *in ovo* with PFOS, indicating that these concentrations are environmentally relevant. These data indicate that immune alterations and brain asymmetry can occur in birds following *in ovo* exposure to environmentally relevant levels of PFOS and warrant further research into the developmental effects of perfluorinated compounds in wildlife.

Molina *et al.* (2006) studied the effects of egg air cell injection of PFOS on the development of White Leghorn chick embryos. In this study, 50 eggs/group were injected with 0.1, 1.0, 10, or 20 μg PFOS/g egg weight before incubation and the effects of PFOS on embryo development were studied. The hatchlings were weighed, examined for gross developmental abnormalities, and raised for 7 days. Chicks were weighed again, following which 20 birds/treatment were randomly chosen for necropsy and major organs (i.e., brain, heart, kidneys, and liver) were removed and weighed. A dose-dependent decrease in hatchability was seen in all treatment groups.

The calculated median lethal dose was 4.9 mg PFOS/g egg. Furthermore, PFOS did not affect post-hatch body or organ weights but did provoke pathological changes in the liver, including bile duct hyperplasia, periportal inflammation, and hepatic cell necrosis at doses as low as 1.0 mg PFOS/g egg. Lastly, PFOS concentrations in the liver increased in a dose-dependent manner. Based on reduced egg hatchability, the LOAEL was determined to be 0.1 mg PFOS/g egg.

A previously published study explored the concentrations, patterns, and temporal trends across three decades (measured in 1983, 1993, and 2003) of 16 perfluorinated alkyl substances (PFASs) in the eggs of the herring gull (*Larus argentatus*) from two geographically isolated colonies in northern Norway (Verreault *et al.* 2007). This study showed that PFOS was the predominant PFAS identified in the eggs, with mean concentrations of up to 42 ng/g (WW) in samples that were collected from 2003. Other PFAS compounds, like perfluorohexane (PFHxS) and perfluorodecane (PFDcS), were found at lower concentrations than was found for PFOS (Verreault *et al.* 2007). Furthermore, the general accumulation profile of perfluorocarboxylates (PFCAs) in herring gull eggs was characterized by high proportions of odd and long carbon (C) chainlength compounds, in which perfluoroundecanoate (C11) and perfluorotridecanoate (C13) were the major species identified, with mean concentrations up to 4.2 and 2.8 ng/g WW, respectively. In both colonies, herring gull egg PFOS concentrations showed an almost 2-fold significant increase from 1983 to 1993, followed by a leveling off up to 2003. The observations from this study suggest that direct and indirect local- and/or remote-sourced inputs (atmospheric and waterborne transport mechanisms) of PFCAs have changed over the past two decades, at least in the context of these two geographically isolated coastal areas of northern Norway.

TOXICITY OF PFOS IN AMPHIBIANS

A summary of the relevant oral toxicological effects of PFOS in amphibian species is described in Table 3.6 and Figure 3.3. The Frog Embryo Teratogenesis Assay-Xenopus (FETAX) was used to determine the possible developmental effects following PFOS exposure of African clawed frogs (*X. laevis*) (Palmer and Krueger 2001, cited in Beach *et al.* 2006). This assay was employed to examine anomalies in the survival, growth, and development of frog embryos and tadpoles following exposure to 0, 1.82, 3.07, 5.19, 8.64, 14.4, or 24.0 mg PFOS/L for 96 hours during early stages of development. Mortality was evident at concentrations > 14.4 mg/L. The LC_{50} value that was determined from these assays was 13.8–17.6 mg/L. It was noted that malformations increased in a dose-dependent manner as PFOS concentration increased. Commonly seen malformations included improper gut coiling, edema, and notochord and facial abnormalities. The minimum concentrations of PFOS that inhibited tadpole growth in two separate assays were 7.97 and 8.64 mg/L. Based on the growth of African clawed frogs, the NOAEL was 5.19 mg/L. The Teratogenic Index (TI), which is defined as the LC_{50} divided by the EC_{50} (both at 96 hours exposure), was in the range 0.9–1.1 when calculated from the results of three separate independent experiments. This TI index shows that there is a low level of risk for potential developmental effects from PFOS exposure in the early life stages of the African clawed

TABLE 3.6
Summary of Relevant Toxicity for PFOS in Amphibian Species

Test Organism	Test Duration	LC_{50} (mg/kg)	NOAEC (mg/L)	LOAEC (mg/L)	Effects Observed at the LOAEC	Reference
Northern leopard frog (*Lithobates pipiens*)	Acute (at week 5) Sub-chronic (at week 16 – partial life cycle)	6.2 (5.12– 7.52)	NA 0.30 mg/L	NA 3.0 mg/L	Based on mortality, and EC_{50} data.	Ankley *et al.* 2004
African clawed frog (*Xenopus laevis*)	96 weeks	15.7 (13.8– 17.6)	4.82	7.97	Based on whole animal growth	Palmer and Krueger 2001

Key: NA = not applicable

PFOS: Health Effects To Amphibians

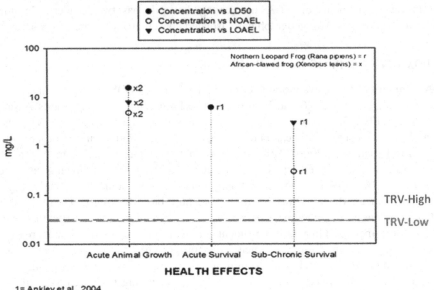

1= Ankley et al., 2004
2= Palmer and Krueger, 2001

FIGURE 3.3   PFOS: Health effects on amphibians.

frog. In addition, the acute LC_{50} value at 96 hours for *X. laevis* embryos was 13.8 mg/L following exposure to the potassium salt (86.7% pure) of PFOS (OECD, 2002).

The effects of PFOS exposure were evaluated in the context of the survival and development of the northern leopard frog (*Lithobates pipiens*) from early embryogenesis through to complete metamorphosis. Northern leopard frogs were exposed to the potassium salt of PFOS (98% fluorosulfonate) at concentrations of 0.03, 0.1, 0.3, 1.0, 3.0, or 10 mg PFOS/L (Ankley *et al.* 2004). Developing tadpoles were exposed to PFOS in glass aquaria (40×20×25 cm) in 10 L of water that was continuously renewed at a flow rate of about 50 mL/min (72 L/d) over the course of a 110-d exposure.

A greater than 90 percent mortality was observed within two weeks of initiation of the study in the 10 mg/L PFOS treatment group. Tadpole survival was unaffected at lower concentrations. Tadpoles readily accumulated PFOS directly from the water. However, time to metamorphosis was delayed and growth reduced in the 3.0 mg/L treatment group. The LC_{50} at week 5 was 6.21 mg PFOS/L No statistically significant effects were noted for tadpoles from treatments <1.0 mg PFOS/L. However, there was a slight increase in time to metamorphosis and a decrease in total length of tadpoles from the 3 mg/L exposure group, and there was a slight increase in the incidence of thyroid follicle cell atrophy.

Overall, the authors suggested that *L. pipiens* was not exceptionally sensitive to PFOS in terms of direct toxicity or bioconcentration potential of the chemical

(Ankley *et al.* 2004). From this study and based on mortality data from the sub-chronic component of the study, a NOAEL level of 0.3 mg/L and a LOAEL level of 3.0 mg/L were determined (Ankley *et al.* 2004).

SUMMARY OF REPTILIAN TOXICITY

No toxicological data for the effects of PFOS on reptiles were obtained from the literature.

TOXICITY REFERENCE VALUES FOR MAMMALS

TRVs FOR INGESTION EXPOSURES FOR THE CLASS MAMMALIA

Based on the information from three species (mouse, rat, and monkey), as described above, the rat was considered to be the mammal most sensitive to sub-chronic oral exposure to PFOS. In a reproductive study, significant adverse effects, following PFOS exposure, included observations of decreased body weights and decreased survival/viability of the rat offspring (Luebker *et al.* 2005b). From these studies, benchmark dose analyses for the TRV-High and TRV-Low derivations in both male and female rats were obtained (Table 3.7) for the selected ingestion TRVs for the class Mammalia. A medium level of confidence was assigned to this TRV because other sub-chronic studies were available, describing additional supporting toxicological end points that included neuro- and immuno-toxicity.

TRV FOR INHALATION EXPOSURE FOR MAMMALIAN SPECIES

TRVs specific to inhalational exposures could not be determined due to the lack of data and weight of evidence for dose-response modeling. Only one definitive study (Rusch *et al.* 1979), and a reference study by OECD (OECD 2002, ATSDR 2015)

TABLE 3.7
Selected Ingestion TRVs for Mammalian Species – PFOS

TRV	Dose (mg/kg-d)	Confidence
TRV-$_{LOW}$	0.0327 (male)	Medium
TRV-$_{HIGH}$	0.0680 (female)	Medium
	0.0521 (male)	
	0.0980 (female)	

Source: Thomford, P.J. 2002. 104–week dietary chronic toxicity and carci-nogenicity study with per-fluorooctane sulfonic acid potassium salt (PFOS); T-6295) in rats. Final Report, 3MT-6295 (Covance Study No 6329-183) Volumes I-IX, 4068 pages. January 2, 2002. 3M St. Paul MN.

TRVs determined by the BMD/BMDL Benchmark Dose approach.

---

were identified. Both studies were acute exposure experimental designs. Data that were derived from a study of an LC_{50} exposure in rats suggested that PFOS absorption can take place by an inhalational exposure route (Rusch et al. 1979). Observed clinical signs from respiratory exposure included emaciation, red material around the nose or nasal discharges, dry rales, and breathing disturbances, or general poor conditioning. In addition, unpublished data (OECD 2002) indicated that exposure of rats to concentrations of PFOS between 1,890 to 54,970 mg/m^3 for 1 hour induced dry rales and other breathing disturbances (ATSDR 2015). Based on these end points, the acute rat inhalation LC_{50} value for PFOS was 5.2 mg/L (ppm) (Rusch et al. 1979).

TRV FOR DERMAL EXPOSURE FOR THE CLASS MAMMALIA

Only one relevant study was identified, in which the acute dermal and ocular irritation test with PFOS was performed in an albino rabbit model (Biesemeier and Harris 1974, as summarized in OECD 2002). This study represents the only mammalian test of dermal or ocular irritancy conducted with PFOS. No sub-chronic or chronic dermal toxicity data were available. The primary dermal irritation score from this study was zero, which indicated no skin irritation potentials or edema from PFOS exposure in this model. Unfortunately, technical details were lacking with regard to the carrier vehicle of PFOS used, the gender of the animals, or of the technical guidelines that were followed (Biesemeier and Harris 1974). In the ocular irritation component of the study, it was determined that PFOS was not an eye irritant.

TOXICITY REFERENCE VALUES FOR AVIAN SPECIES

Based on the data from definitive chronic exposure studies conducted on northern bobwhite andmallard, as described above, both NOAEL and LOAEL TRVs were derived from chronic oral reproductive toxicity studies reported by Newsted et al. (2007). Based on observations of only slight effects being observed in mallard that were fed 10 mg PFOS/kg feed, bird populations that were exposed to dietary concentrations of PFOS ≤ 6.4 mg PFOS/kg feed would not be expected to experience adverse effects on the survival of adults or offspring, on the reproductive performance of the adults, or the survival, growth, or development of the offspring (Newsted et al. 2007).

From these reproduction studies, the LOAEL for avian species was 6.40 mg PFOS/kg BW-d, whereas the NOAEL was 1.50 PFOS/kg BW-d. Therefore, derivation of a NOAEL-based TRV from this study was used with the application of an uncertainty factor (UF) of 10 to account for interspecific variability and in the extrapolation from the chronic studies (UF of 1) for a TRV of 0.15 mg/kg-d (Table 3.8). A low-to-medium level of confidence was assigned to this TRV because other chronic studies in the bobwhite (Gallagher et al. 2003a–c) were available that indicated decreased 14-day survival of adult male and female northern bobwhite. The derivation of a LOAEL-based TRV from this single study was used with the application of a UF of 10 to account for interspecific variability; in the extrapolation from the described chronic study, a LOAEL-based TRV (UF of 1) of 0.64 mg/kg-d,

TABLE 3.8
Selected Ingestion TRVs for Avian Species – PFOS

TRV	Dose (mg/kg-d)	Confidence
TRV_{-LOW}	0.15	Low - Medium
TRV_{-HIGH}	0.64	Low - Medium

Sources:

Newsted, J.L., Coady, K.K., Beach, S.A., Butenhoff. J. L., Gallagher, S., Giesy, J.P. 2007. Effects of perfluorooctane sulfonate on mallard and northern bobwhite quail exposed chronically via the diet. Environ.Toxicol. Pharmacol. 23: 1–9.

Gallagher, S.P., Van Hoven, R.L., Beavers, J.B., Jaber, M. 2003a. PFOS: A reproduction study with northern bobwhite. Final report. Project No. 454-108. USEPA Administrative Record AR-226-1831. Wildlife International, Ltd., Easton, MD, USA.

Gallagher, S.P., Van Hoven, R.L., Beavers, J.B., Jaber, M. 2003b. PFOS: A reproduction study with mallard. Final report. Project No. 454-109. USEPA Administrative Record AR-226-1836. Wildlife International, Ltd., Easton, MD, USA.

Gallagher, S.P., Van Hoven, R.L., Beavers, J.B., 2003c. PFOS: a pilot reproductive study with the northern bobwhite. Wildlife International, Ltd., Project No. 454-104. USEPA Administrative Record AR-226-1817

TRVs determined by the NOAEL/LOAEL approach.

also with a low-to-medium confidence level, was assigned (see Table 3.8). A low-to-medium level of confidence was applied because chronic developmental studies conducted in the quail showed evidence of clinical signs, lethargy, reduced reaction to external stimuli, and a ruffled appearance starting at week 5 (Newsted *et al.* 2007). However, no apparent treatment-related effects were found on body weight for northern bobwhite and mallard exposed to 10 mg/kg feed but body weight was reduced in the higher dose groups in both species.

TOXICITY REFERENCE VALUES FOR AMPHIBIANS

The FETAX assay assisted in determining possible developmental effects from acute PFOS exposure of African clawed frogs (*X. laevis*) (Palmer and Krueger 2001, cited in Beach *et al.* 2006). This assay was employed to examine anomalies in the survival, growth, and development of frog embryos and tadpoles during the early stages of development. Mortality was evident at concentrations > 14.4 mg/L. The LC_{50} value was calculated as 13.8–17.6 mg/L. It was noted that malformations increased in a dose-dependent manner as PFOS concentration increased, as described above.

Due in part to the limited number of studies conducted on amphibian species, the selection of this acute exposure study for derivation of TRVs, and the fact that this single study evaluated only one species, there is a low level of confidence given to these TRVs. Additionally, data from studies conducted on the northern leopard frog suggests that this native species is more sensitive than *Xenopus* (Ankley *et al.* 2004). From this study, and based on mortality data from the sub-chronic component of the

TABLE 3.9

Selected TRVs for Amphibians – PFOS

TRV	Concentration (mg/L/day)	Confidence
TRV-LOW	0.003	Low
TRV- HIGH	0.075	Low

Source: Ankley, G.T., D.W. Kuehl, M.D. Kahl, K.M. Jensen, B.C. Butterworth and J.W. Nichols 2004. Partial life-cycle toxicity and bioconcentration modeling of perfluorooctane sulfonate in the northern leopard frog (*Lithobates pipiens*). Environ. Toxicol. Chem. 23: 2745–2755.
TRVs determined by the NOAEL/LOAEL approach.

study, a NOAEL level of 0.3 mg/L and a LOAEL level of 3.0 mg/L were determined (Ankley *et al.* 2004). A summary of the selected TRVs for amphibians is described in Table 3.9.

From these studies, derivation of a NOAEL-based TRV from this one study applied an uncertainty factor (UF) of 100 to account for interspecific variability (UF of 10) and in the extrapolation from a single sub-chronic study (UF of 10) for a TRV of 0.003 mg/L (Table 3.9). A low level of confidence was assigned to this TRV because of the limited subsets of data available, with relatively few data supporting the derived TRVs. The derivation of a LOAEL-based TRV from this one study applied a UF of 40 to account for interspecific variability (UF of 10) and, in the extrapolation from the described sub-chronic study, a LOAEL-based TRV (UF of 4) of 0.075 mg/L was derived, also with low confidence.

TOXICITY REFERENCE VALUES FOR REPTILES.

No toxicological data for the effects of PFOS on reptiles were located during the literature search.

IMPORTANT RESEARCH NEEDS

There are considerable toxicity data on mammals and of the human health effects in response to PFOS exposure, as described in some detail above. The available toxicity data on wildlife, however, is largely limited to studies conducted on mammals and birds. There is a paucity of data on the response of reptiles to PFOS exposure and little data on the response of amphibians to acute oral PFOS exposure. Few details have been published with regard to the acute and chronic health effects of inhalational exposure of mammals to PFOS, and more detailed chronic exposure studies of mammals, and particularly birds, amphibians, and reptiles, are urgently needed, due to the relative lack of technical details available on published reports on the exposure to PFOS in the case of birds, and the absence of high-quality exposure studies in the case of reptiles and amphibians.

From an ecological perspective, these missing data are important for understanding the relevance of environmental persistence, the ubiquitous presence of PFOS in water columns and soils, and an appreciation that these groups of animals are likely to be exposed to PFOS at and away from contaminated sites as a result of long-range transport. Clearly, further PFOS toxicological studies, with a particular focus on birds, amphibians, and reptiles, are warranted, and data derived from these studies may enable the derivation of TRVs for non-mammalian but important ecological receptors.

4 Perfluorohexane Sulfonate (PFHxS)

Allison M. Narrizano

Perfluorohexane sulfonate (PFHxS) is a six-carbon per-polyfluorinated alkyl substance (PFAS). The elimination half-life of PFHxS is about 1 month in mice and 4 months in monkeys, depending on sex (Sundstrom *et al.* 2012), with females eliminating the chemical more quickly. In rats exposed to PFHxS, the differences in elimination half-life are dramatic (i.e., 2 days in females and 29 days in males). No acute toxicity data in rodents were found for PFHxS..

The National Toxicology Program (NTP) conducted 28-day repeated oral gavage studies with male and female rats exposed to various PFSAs. PFHxS was delivered once daily to males at 0–10 mg/kg-d and once daily to females at 0–50 mg/kg-d; higher doses were administered to females based on prior identification of a higher maximum tolerated dose in females and supporting kinetic and toxicity information from the literature. No treatment-related mortality occurred in animals exposed to PFHxS (NTP 2019b). Liver weights were higher in rats exposed to PFHxS. Gender differences in plasma chemical concentrations were observed in PFHxS-exposed animals; males exposed to PFHxS had decreased concentrations of cholesterol, triglycerides, and globulin. As with other parameters evaluated in animals exposed to PFHxS, sex differences were observed in terms of gene expression; compared with their respective controls, exposed males exhibited increased peroxisome proliferator-activated receptor-alpha (PPARα) and constitutive androstane receptor (CAR) activities, whereas exposed females presented increased CAR but not PPARα activities. Estrous cyclicity and sperm parameters were unaltered in rats exposed to PFHxS. Exposure of male and female rats to PFHxS reduced levels of thyroid hormones (e.g., triiodothyronine (T3) and thyroxine (T4)), but had no effects on concentrations of thyroid-stimulating hormone (TSH) or testosterone. Organs of animals exposed to PFHxS appeared normal following histopathological examination.

In another study, parental (P) generation rats were exposed to PFHxS (at 0, 0.3, 1, 3, or 10 mg/kg-d) *via* oral gavage for 14 days prior to mating, during mating, and until the day before sacrifice (lactation day 21 in females or after 42 days of treatment in males). F_1 generation animals were not directly exposed to PFHxS (i.e., they were only exposed *in utero* and *via* lactation). No reproductive or developmental effects were observed, and no treatment-related effects occurred in dams or offspring (Butenhoff *et al.* 2009). PFHxS-exposed P generation males exhibited decreased cholesterol levels and increased prothrombin concentrations. Additionally, P generation males exposed to 3 or 10 mg/kg-d had increased liver weights, hepatocellular hypertrophy, hyperplasia of thyroid follicular cells, and decreased hematocrit percentage levels. Finally, high-dose males had decreased concentrations of triglycerides, and some increases in clinical chemistry parameters.

The same P generation dose regimen was used in a study with mice exposed to PFHxS (0, 0.3, 1, or 3 mg/kg-d) *via* oral gavage. Pup-born-to-implant ratio was unaffected by PFHxS, though there was an equivocal finding of slightly decreased live litter size at 1 and 3 mg/kg-d (Chang *et al.* 2018). As with other studies, PFHxS-exposed P generation males displayed increased hepatocellular hypertrophy, decreased serum cholesterol level, and increased alkaline phosphatase activity. Otherwise, there were no toxicologically significant effects on reproduction, hematology, clinical chemistry parameters, neurobehavior, or histopathology, and no significant treatment-related effects on post-natal survival, development, or date of onset of puberty. Data from a reproductive and developmental toxicity study with deer mice (*Peromyscus maniculatus*) exposed to PFHxS supported findings from studies on laboratory rodents and suggest that PFHxS is less toxic than PFOS (Narizzano and Quinn, unpublished data).

The effect of a single dose delivered *via* oral gavage of PFHxS (0, 0.61, 6.1, or 9.2 mg/kg) at post-natal day 10 (PND10) on brain development was evaluated in a study with male and female mice. At 2 and 4 months of age, mice receiving the highest dose showed significantly reduced locomotion, rearing, and total activity during the first 20 minutes of testing and significantly increased activity during the last 20 minutes of testing (Viberg *et al.* 2013).

5 Perfluororoheptanoic Acid (PFHpA)

Michael J. Quinn Jr.

CONTENTS

Perfluoroheptanoic acid (PFHpA) is a seven-carbon per- or polyfluorinated alkyl substance (PFAS), that is a metabolite of longer- chained carbon compounds. Perfluorinated heptanoic acid is a fluoroalkanoic acid. Bioaccumulation is not expected to be significant, as the half lives in male and female rats were calculated to be 0.10 and 0.05 days, respectively (Ohmori *et al*. 2003).

MAMMALIAN TOXICITY

Very few mammalian toxicity studies with PFHpA could be located in the literature. One study investigated the *in vitro* toxicity of PFAS, individually or in combination, including PFHpA, to a human liver cell line (HepG2; Ojo *et al*. 2020). By itself, PFHpA was the least toxic of the perfluoroalkyl carboxylic acids (compared with PFDA, PFNA, and PFOA). A synergistic effect on cytotoxicity was observed with a combined exposure to PFHpA and PFOS, and an antagonistic interaction was found when PFHpA was combined with PFOA. In the absence of *in vivo* animal test data, some human epidemiological studies were conducted that might contribute to the body of mammalian toxicology literature for this chemical. No correlations were found between mean PFHpA and testosterone levels in 105 men from low- and high-testosterone test groups (Joensen *et al*. 2009). Although median levels of serum PFHpA were the same in the two groups, a significant difference in serum PFHpA levels was reported between children with or without asthma (Dong *et al*. 2013), with 53.3% and 70.6% of non-asthmatic and asthmatic children, respectively, having detectable serum concentrations of PHFpA.

In the one oral exposure rat study that could be located, PFHpA had no effect on the accumulation of triglycerides and the induction of peroxisomal beta-oxidation in the liver of male and female rats (Kudo and Kawashima 2003). In the one dermal exposure study that could be located, Sprague Dawley rats were dermally exposed to 250 or 1,000 mg/kg PFHpA for two weeks (Han *et al*. 2020). At 1,000 mg/kg, 83% of the rats died with severe ulcerative dermatitis at the exposure site. Renal tubular necrosis, hepatocellular necrosis, and germ cell degeneration were observed in both low- and high-PFHpA treatment groups.

AMPHIBIAN TOXICITY

In a frog embryo teratogenicity assay, *Xenopus* embryos were exposed to doses of PFHpA from 0.25 to 1.25 mM (Kim *et al.* 2015). The median lethal concentration (LC_{50}) was determined to be 942.4 µM, and PFHpA was identified as a potential teratogen and developmental toxicant. Developmental abnormalities included gut mis-coiling, stunted body, multiple edemas, microcephaly, skeletal kinking, and shorter body length. Although the dose at which most of these effects were observed was not reported, 24% tadpoles that were treated with 1,000 µM of PFHpA experienced whole body length reductions. Tadpoles exposed to 130 µM PFHpA had enlarged livers at stage 36. Additionally, at this exposure level, enlarged atria, a loss of the atrial septum, and thinner atrial and ventricular walls were also observed.

AVIAN TOXICITY

Adult northern bobwhite were exposed to drinking water containing nominal concentrations of 20 ng/mL, 1 ng/mL, or 0.1 ng/mL of PFHpA over 90 days (Thompson 2018). Following egg laying, juvenile F_1 offspring from the P generation were assessed for 30 days and mated. Residue analysis was conducted on adult and juvenile liver tissue as well as on eggs. PFHpA did not affect hatching success, reproduction, or adult or juvenile survival at 1.860 or 1.745 µg/kg-day, respectively. Residue analysis showed that female birds reduced their body burden of PFHpA through deposition into eggs. The authors noted that patterns of liver damage were identified in adult birds following chronic exposure, although any differences between treatment groups were not significant.

6 Perfluorononanoic Acid (PFNA)

Allison M. Narizzano

CONTENTS

MAMMALIAN TOXICITY

Perfluorononanoic acid (PFNA) is a nine-carbon per- polyfluorinated alkyl substance (PFAS). Male and female rats cleared PFNA in 30 and 2.44 days, respectively (Ohmori *et al.* 2003). No acute toxicity data in rodents were identified for PFNA.

The National Toxicology Program (NTP) conducted 28-day repeated dose oral gavage studies with male and female rats exposed to PFNA, delivered once daily in the form of one of five doses to males in the range 0–10 mg/kg-d and females at 0–25 mg/kg-d (0, 0.63, 1.25, 2.5, 5, or 10, and 0, 1.56, 3.12, 6.25, 12.5, or 25 mg/kg-d for males and females, respectively). Higher doses were administered to females based on prior identification of a higher maximum tolerated dose and supporting kinetic and toxicity information from the literature. Dose-dependent mortality and reduced body weights occurred in males and females exposed to PFNA (NTP 2019). Males had reduced survival compared to the no-PFNA controls at 5 mg/kg-d and above, and females at 12.5 mg/kg-d and above. In animals that survived 28 days of dosing, liver weights and liver enzyme activities were increased by exposure to PFNA. Male and female rats administered PFNA had increased expression of the peroxisomal acyl-coenzyme A oxidase 1 gene, *Acox1,* and the cytochrome P450 genes, *Cyp4a1, Cyp2b1, and Cyp2b2,* compared with controls, suggesting increased peroxisome proliferator-activated receptor alpha (PPARα) and constitutive androstane receptor (CAR) activity (NTP 2019). Estrous cyclicity results were inconclusive in animals exposed to PFNA. Sperm counts were reduced following exposure to PFNA. Males exposed to PFNA had decreased thyroid-stimulating hormone (TSH) levels, whereas females exposed to PFNA had increased and unaltered levels of TSH, respectively. Plasma concentrations of testosterone were decreased in PFNA-exposed male rats and increased in female rats exposed to PFNA. In rats exposed to PFNA, the liver and bone marrow were target organs, as reflected histopathologically by hepatocyte cytoplasmic alteration, hepatocyte hypertrophy, and bone marrow hypocellularity. Additionally, animals exposed to PFNA displayed splenic and thymic atrophy.

Wolf *et al.* (2010) conducted a study in which timed pregnant mice (wild-type (WT) and PPARα-null mice) were exposed to PFNA (0, 0.83, 1.1, 1.5, or 2 mg/kg-d) *via* oral gavage at gestation date (GD) GD1-18. In both strains of mice, maternal weight gain, number of implantations, litter size, and pup weight at birth were unaffected by PFNA (Wolf *et al.* 2010). In WT mice, exposure to 1.1 or 2 mg/kg-d reduced the number of live pups at birth and the percentage survival of offspring to weaning. Additionally, eye opening was delayed and pup weight at weaning was reduced in WT pups from the 2 mg/kg-d dose group. The number of live pups, pup survival, pup body weight gain, and time to developmental landmarks did not differ between PPARα-null mice exposed to PFNA and no-PFNA controls, suggesting that PPARα expression favors the PFNA-induced developmental toxicity in mice. As with other studies with PFNA, WT dams and pups demonstrated dose-dependent increases in liver weights, but effects were slight in PPARα-null mice.

Das *et al.* exposed timed pregnant mice to PFNA (0, 1, 3, 5, or 10 mg/kg-d) *via* oral gavage at GD1-17. Dams exposed to 10 mg/kg-d exhibited overt toxicity (i.e., maternal weight loss) and were sacrificed on GD13 (Das *et al.* 2015). Of these animals, 100% of pregnant animals that were examined displayed full litter resorption. In contrast, dams in other PFNA dose groups gained weight as expected, and there were no changes in the number of implantations, number of live fetuses, or fetal weights, compared with the controls. As with other perfluoroalkyl acids exposure of P and F_1 generation mice to PFNA was associated with increased liver weights; however, the pattern of neonatal mortality differed slightly from what was seen with perfluorooctanesulfonic acid (PFOS) and perfluorooctanoic acid (PFOA). Specifically, although dose-dependent neonatal mortality occurred, the majority of PFNA-exposed pups survived longer after birth than did those exposed to PFOS or PFOA. Surviving neonates exposed to PFNA exhibited dose-dependent decreases in body weight gain and delays in eye opening and the onset of puberty.

In a recent study by Narrizano and Quinn (unpublished data), PFNA did not markedly affect reproduction or development in white-footed mice.

Immune responses were reduced in male rats exposed orally to PFNA (0, 1, 3, or 5 mg/kg-d) for 14 days. Specifically, thymus weights, and the concentration of interleukin IL-2 were reduced in animals exposed to 3 or 5 mg/kg-d, whereas interleukin IL-1 concentration was increased (Fang *et al.* 2009). Furthermore, spleen weights were reduced, and the number of apoptotic cells were increased in a dose-dependent manner (Fang *et al.* 2010). PFNA also modulated CD4+CD8+ thymocytes, which suggests that PFNA may interfere with the process of thymocyte maturation and differentiation (Fang *et al.* 2008). In contrast, innate immune cells seemed to be the cellular target of PFNA in the spleen, as the numbers of F4/80+, CD11c+, and CD49b+ thymocytes decreased in response to exposure to PFNA.

TOXICITY TO OTHER ANIMAL CLASSES

No toxicity information was found on effects from exposures to PFNA to birds, amphibians, or reptiles.

7 Perfluorobutane Sulfonate (PFBS)

Marc A. Williams

CONTENTS

INTRODUCTION

Perfluorobutane sulfonate (PFBS) is a four-carbon (C4), fully fluorinated alkane that belongs to a growing number of per- and polyfluorinated alkyl substance (PFAS) compounds, that are synthesized for use in many different everyday consumer products with the intent of making them stain-, grease-, and water-resistant. Like many perfluoroalkyl carboxylates (PFCAs) and perfluoroalkane sulfonates (PFSAs), PFBS exhibits many valuable physicochemical properties, including its ability to reduce surface tension, its stability, and its hydrophobic properties. Thus, PFBS has been used in cleaning agents, paint formulations, and water-impermeable products, and to provide water-proofing for clothing and mattresses, to prevent food from sticking to cookware, and to resist staining of carpeting and furniture upholstery (Rosal *et al.* 2010).

PFBS was synthesized originally as a safer alternative to perfluorooctanesulfo-nate (PFOS), as a result, in large part, to increased awareness in the early 2000s of the prolonged environmental persistence and the potential for bioaccumulation of longer-chain (C8) PFAS compounds, like PFOS. Thus, new industrial markets appeared for shorter-chained (C4) PFAS compounds, including PFBS, and interest in PFBS increased. Short-chain C4 compounds, like PFBS, were considered to be less bioaccumulative and less toxic than PFOS due to their shorter chain length. This conferred PFBS with greater solubility and more rapid elimination from living organisms as compared to PFOS or similar long-chain C8 compounds, and with reduced affinity for critical biological receptors, like peroxisome proliferator-activated receptor-alpha (Renner 2006, Olsen *et al.* 2009, Lieder *et al.* 2009).

Nonetheless, the increased use of the shorter-chain PFBS as an alternative to longer-chain PFASs has led to PFBS leaving an environmental contamination footprint across both aquatic and terrestrial media. The challenge is that PFBS can be released directly to the environment or it can be formed through the degradation of perfluo-robutanesulfonyl fluoride-derived compounds (Newsted *et al.* 2008). Consequently, there are environmental health impact concerns with respect to not only materials manufactured with PFBS as a key component, but also from degradation of PFBSF-derived industrial products (Olsen *et al.* 2009).

PFBS has been found in food contact materials, dust, fire-fighting foams, and drinking water supply chains. Like many other PFAS compounds, plausible pathways of exposure to PFBS include oral exposure from food and water, inhalational exposure, and possible dermal contact, with the oral exposure route being considered to be the primary pathway.

ENVIRONMENTAL FATE AND TRANSPORT

PFBS does not occur naturally in the environment. Through its manufacture and use, PFBS is released into environmental media and substrates *via* many waste streams. The major environmental concern is the inherent resistance of PFBS to hydrolysis, photolysis, and biodegradation, processes which collectively reduce environmental persistence, so that resistance to these processes results in increased persistence (Sundström *et al.* 2012). Many sampled sites in the USA have identified PFBS in surface waters across a number of states, including North Carolina, Georgia, and New Jersey, as well as the Upper Mississippi River Basin (Post *et al.* 2013, Lasier *et al.* 2011, Nakayama *et al.* 2007, 2010); and in many wastewater, marine water, sewage treatment facility effluents, and soil samples (Houtz *et al.* 2016, Zhao *et al.* 2012, Sepulvado *et al.* 2001). Environmental contamination with PFBS was also reported following the use of aqueous film-forming foams (AFFF) during fire-training exercises at Wurtsworth Air Force Base, MI, USA (ASTSWMO 2015).

Once PFAS are released into the environment, the two main modes of transport are atmospheric (Taniyasu *et al.* 2013) and aquatic transmission (Armitage *et al.*

2009). The aquatic mode of transport is the one of greatest concern, due, in large part, to the relatively high polarity and solubility of ionic PFASs (Armitage *et al.* 2009). On release to the atmosphere, the estimated vapor pressure of 2.68×10^{-2} mm Hg for PFBS informs us that it is likely to remain as an atmospheric vapor (HSDB 2016). As a vapor-phase compound, PFBS is degraded in the atmosphere, following its reaction with photochemically synthesized hydroxyl radicals, with a half-life for this reaction of about 115 days (HSDB 2016). Since PFBS lacks reactive chromophores that absorb at wavelengths greater than 290 nm, it is unlikely that PFBS will be susceptible to direct photolysis on exposure to sunlight.

With an estimated soil adsorption coefficient (K_{OC})of 180, PFBS is expected to display moderate mobility if released into the soil (HSDB 2016). PFBS has an estimated pK_a (negative $\log_{10}$ of the acid dissociation constant, K_a) of -3.31, which indicates that PFBS will exist in the environment in anionic form, which will not adsorb as strongly to soils, comprising organic carbon and clay, as will their neutral counterparts. If released into water columns, the estimated K_{OC} of 180 indicates that PFBS is expected to adsorb to suspended solids and sediments. According to its very low estimated pK_a value (HSDB 2016), volatilization of PFBS from water surfaces is also not expected,. The pK_a value, -3.31, also indicates that PFBS is unlikely to volatilize from moist soil surfaces. PFBS also lacks functional groups that hydrolyze under environmental conditions at a pH of 5 to 9. Thus, hydrolysis would not be expected to be an important environmental fate process (HSDB 2016).

In terms of the biodegradation of PFBS, studies by Quinete *et al.* (2010) sourced an inoculum from the Rhine River, and, in a closed bottle test (i.e., OECD 301D), that was incubated for 28 days at 20°C, found that less than three percent of the original PFBS (starting at 73 mg/L) had biodegraded. Also, when applying a manometric respirometer test (OECD 301F) with activated sludge, less than one percent of the original PFBS (starting at 100 mg/L) was biodegraded over a 40-day period (Quinete *et al.* 2010). In a fixed-bed bioreactor, this research group showed that, on testing an inoculum of the Rhine River sample, PFBS was observed to be resistant to biodegradation, conducted at room temperature and under dark conditions, over a 28-day period (Quinete *et al.* 2010).

In the context of aquatic toxicology, a multi-generational test was performed with a freshwater macroinvertebrate species, *Chironomus riparius* (Diptera: Chironomidae), in an attempt to determine the long-term generational effects of exposure to PFBS and other PFAS compounds (Marziali *et al.* 2019). Interest in designing this study stems from the knowledge that PFASs are ubiquitously found in aquatic ecosystems. This study exposed *C. riparius* larvae for ten generations to 10 μg/L nominal concentrations of PFASs (including PFBS) – a concentration that corresponded to levels found in European water columns. Although no effects were found on survival, development, and reproduction, all PFASs tested showed reduced growth for most, or across several generations; however, potential effects at the population level were not formally demonstrated in this study, and the risk of toxicity in "real-world" ecosystems appears unlikely (Marziali *et al.* 2019).

An ecotoxicological study by Gebbink *et al.* (2016) found that the emerging PFASs, such as PFBS and F-35B (a chlorinated polyfluorinated ether sulfonic acid), were detected for the first time in Arctic (Greenland) wildlife, although it was found that the levels were up to four orders of magnitude lower than those of PFOS. PFBS was detected in all polar bear livers tested, at mean ($\pm$ standard error, SE) concentrations of 0.032 ± 0.008 ng/g and in approximately 67 percent of killer whale livers tested at concentrations of 0.0052 ± 0.0017 ng/g, but the PFBS concentration was below the threshold of detection in ringed seal livers (Gebbink et al. 2016). The presence of PFBS in arctic wildlife follows the observation that PFBS was reported in sediment and water samples collected from the Arctic region (Stock et al. 2007, Yamashita et al. 2008). It is thought that PFBS transportation pathways in the Arctic would include oceanic transportation of PFBS, atmospheric transport of volatile PFBS precursors, and latent degradation of other PFAS to PFBS (Gebbink et al. 2016).

BIOACCUMULATION AND ELIMINATION

In arguably the first study to explore the toxicokinetics of PFBS in mammals (Olsen et al. 2009), the elimination pharmacokinetics of PFBS were studied in rats, humans, and macaque monkeys. PFBS was distributed largely in the extracellular space. After the initial phase of the whole-body elimination of PFBS that was observed by Olsen et al. (2009), profiling of the early phases supported a multiple compartment model for the distribution and elimination of PFBS (Olsen et al. 2009), although the mechanism is still poorly understood. Others have speculated that the organic anion transporting polypeptide, Oatp1, plays a key role in the reabsorption of the PFBS congener, PFOA, in rat proximal tubules (Kudo and Kawashima 2003). However, the extent to which perfluoroalkylsulfonate kinetics is determined by organic anion transporter-facilitated processes, and how these might differ between species and genders within a particular species is still poorly understood (Olsen et al. 2009).

Compared with PFOS (as a typical example of a long-carbon-chain compound), which has a serum elimination half-life estimated at about 7 days in rats (Johnson et al. 1979), and a mean ($\pm$ SE) of 132 ± 13 days in male monkeys, and 110 ± 26 days in female macaques (Noker and Gorman 2003), PFBS is expected to display a more rapid serum elimination rate in these species due to its shorter perfluorinated chain and its probable higher percentage of free serum-borne PFBS concentration (Kerstner-Wood et al. 2003). Species-dependent differences in serum elimination rates across the aforementioned test species are thought to be guided by a saturation-specific renal resorption process (Andersen et al. 2006).

The toxicokinetics of PFBS were evaluated in Sprague Dawley rats and crab-eating macaques (Macaca fascicularis) (Chengelis et al. 2009). In this study, three male and three female animals macaques (each approximately three years of age) were each dosed once at 10 mg/kg (5 mL/kg) in the form of a single intravenous bolus injection, following which the animals were monitored for seven days (Chengelis et al. 2009). In a single-dose study in rats (approximately seven weeks of age), 12 male and 12 female rats were each dosed once at 10 mg/kg (5 mL/kg) in the form of a single intravenous bolus injection. It was found that the male monkeys had a higher exposure to PFBS, and a longer elimination half-life than did the female monkeys.

However, these data needed to be interpreted with caution, since the mean values were influenced by one male monkey with a markedly higher-than-normal plasma PFBS concentration; thus, no major gender differences were seen in this study, when correcting for that outlier (Chengelis *et al.* 2009). From the rat studies, exposures to PFBS were approximately 7–8-fold higher for male rats than with female rats (Chengelis *et al.* 2009). Furthermore, the terminal serological half-life for PFBS in female rats (0.64 h) was approximately 3-fold lower than for the males (2.1 h). In addition, apparent serological clearance of PFBS was approximately 7–8-fold higher for female rats than for male rats (Chengelis *et al.* 2009).

In the study by Olsen *et al.* (2009), the comparative elimination pharmacokinetics of PFBS were studied in rats, macaques, and humans. This paper described intravenous elimination studies in rats and monkeys, and oral uptake and elimination studies only in the rat (Olsen *et al.* 2009). It was found that urine was the major route of elimination in both species.

The mean ($\pm$ SE) terminal serum PFBS elimination half-lives in rats, after intravenous (i.v.) administration of 30 mg/kg PFBS, were 4.51 ± 2.22 hours and 3.96 ± 0.21 hours for male and female rats, respectively. In monkeys, the mean terminal serum PFBS elimination half-lives, after i.v. administration of 10 mg/kg PFBS, were 95.2 ± 27.1 hours and 83.2 ± 41.9 hours for males and females, respectively. Although the terminal serum half-lives in male and female rats were similar, clearance (CL) was greater in female rats (469 ± 40 mL/h) than male rats (119 ± 34 mL/h). Thus, both males and females excreted about one-half to three-quarters of a single administered oral (rats) or intravenous dose of PFBS (in rats and monkeys) during the initial 24-h period following exposure (Olsen *et al.* 2009).

Similarly, rat serum elimination rates of PFBS were markedly faster than those of PFOS, where a serological elimination half-life of PFBS in rats was only 4.5 hours (Olsen *et al.* 2006), compared with a much longer whole-body elimination half-life for PFOS of approximately 100 days (Johnson *et al.* 1979). This study explored the elimination of ^{14}C from ^{14}C-labeled potassium salt of PFOS (K+PFOS) in the urine and feces of six male rats following a single i.v. dose of 10.8 mg/kg (Johnson *et al.* 1979).

Studies on the distribution of ^{35}S-labeled PFBS following dietary exposure (16 mg/kg-d) in adult male C57BL/6 mice for 1, 3, or 5 d, revealed the distribution of PFBS across all 20 tissues/organs studied, which illustrated the ability of PFBS to exit the peripheral blood system and enter tissues (Bogdanska *et al.* 2014). Highest concentrations were found in the liver and kidney, gastrointestinal tract, peripheral blood, whole bone and cartilage, the lung, and the thyroid gland. In addition, relatively high levels were found in the male reproductive organs, with the notable exception of the testis (Bogdanska *et al.* 2014). However, when compared with PFOS, exposure to PFBS resulted in a 5–40-fold lower concentrations of PFBS in the tissues examined, with lower concentrations in the liver and lung relative to those found in the blood (Bogdanska *et al.* 2014).

More recently, a comparative toxicokinetic analysis was reported for several PFAS compounds (including PFBS) in male and female Sprague Dawley rats following exposure to PFASs by a single intravenous (4 mg/kg) or oral gavage route (4, 20, or 100 mg/kg) of delivery (Huang *et al.* 2019). The group reported a plasma half-life of PFBS after gavage administration of 3.3 hours in males and 1.3 hours in females, compared with PFOS, which exhibited a plasma half-life of about 20 days in both males and female rats (Huang *et al.* 2019).

Importantly, shorter-chain PFAS compounds, like PFBS, exhibit increased solubility compared with their longer-chain counterparts, like PFOS, which contributes to the faster absorption, lower observed tissue distribution, and higher elimination rates of the former, which leads to markedly decreased systemic exposures (Huang *et al.* 2019). Following oral gavage administration, it was found that PFBS (like the other PFASs tested) was absorbed and distributed into the tissues within 24 hours. PFBS (like the other PFASs tested) was found in the liver, kidney, and brain, with the greatest concentrations found in the liver, followed by the kidney (Huang *et al.* 2019). These observations were consistent with previously discovered PFAS pharmacokinetics, including those for PFBS (Bogdanska *et al.* 2014, Sundström *et al.* 2012, Chang *et al.* 2012).

MAMMALIAN ORAL TOXICITY

Although many toxicological studies have investigated the toxicity of the potassium or ammonium salts of PFBS, these will not be discussed in the current wildlife toxicity assessment (WTA). The intent of this WTA is to summarize and discuss relevant research data derived from studies of the free acid form of PFBS only. Only a few acute or sub-chronic/chronic studies have been conducted on the effects of PFBS as the free acid, compared with the anionic forms of PFBS. The toxicology of PFBS salts should be surveyed in future and separate toxicological assessments carried out.

MAMMALIAN ORAL TOXICITY – ACUTE/SUB-ACUTE

According to the USACHPPM Technical Guidance 254 (USACHPPM 2000), acute exposures are defined as single exposures occurring in a single day and subacute exposures are repetitive exposures of less than 14 days. In one of the earliest studies, conducted at Bayer AG, Germany (Bomhard and Loser 1996), five groups, each consisting of ten young male adult rats (average weight: 174 g), were dosed at 50, 100, 300, 600, or 800 µl/kg, although the exact concentrations and mode of oral delivery of PFBS were not supplied in the text (Bomhard and Loser, 1996). The rats showed corresponding mortality rates of 0, 20, 60, 80, and 100%, respectively.

Clinical signs in rats included poor overall condition, sedation, piloerection, bloody muzzle, loss of body weight, and dyspnea. Symptom onset was observed as early as 30 min post exposure and, by day 14, some animals showed persistent signs of toxicological effect. Deaths were seen from day 1 through to day 7, and, on necropsy, for those animals surviving to day 14, evidence of mushy stomach contents, and highly reddened lung parenchyma was visible (Bomhard and Loser 1996). These data permitted estimation of an acute oral median lethal dose (LD_{50}) for male rats of 236 µl/kg, which corresponded to 430 mg/kg, with a 95 percent confidence interval of 153–345 µl/kg (Table 7.1; Bomhard and Loser 1996).

TABLE 7.1

Summary of Acute Oral Toxicity for PFBS (Free Acid) in Mammals

Test Organism	LD$_{50}$ (mg/kg)	NOAEL (mg/kg-d)	LOAEL (mg/kg-d)	Effects Observed at the LOAEL	Reference
			Test Results		
Rats (male)	430	ND	ND	LD$_{50}$ mortality rates of 0, 20, 60, 80 and 100 percent at 50, 100, 300, 600, and 800 µl/kg PFBS, respectively	Bomhard and Loser (1996)

ND – not determined

MAMMALIAN ORAL TOXICITY – SUB-CHRONIC

No studies were found in the literature that explored sub-chronic oral toxicity, reproductive toxicity, or developmental toxicity of the PFBS free acid.

MAMMALIAN ORAL TOXICITY – CHRONIC

No published studies were identified that explored the chronic oral toxicity of the PFBS free acid. However, from a sub-chronic duration oral gavage study conducted in Sprague Dawley rats (Lieder *et al.* 2009), in which animals were exposed to the potassium salt of PFBS (K+PFBS), the United States Environmental Protection Agency (USEPA) derived a provisional chronic reference dose (p-RfD) for PFBS as the free acid (see below and USEPA 2014) for more information). In the study (Lieder *et al.* 2009), K+PFBS was administered daily to each of ten Sprague Dawley rats/sex/dose *via* an oral gavage procedure for approximately 90 days at doses of 0, 60, 200, or 600 mg/kg-d. No treatment-related effects were noted in terms of food consumption, mortality, or behavior. No changes to the body weight of either male or female rats were seen for any dose groups throughout the treatment period (Lieder *et al.* 2009). In male rats, but not females, there was a statistically significant decrease in absolute and relative spleen weights (spleen-to-body-weight ratio) in the 60, 200, and 600 mg/kg-d groups, compared with the 0 mg/kg-d controls (Lieder *et al.* 2009).

The average total protein and albumin concentrations were decreased in female rats but not in males in the 600 mg K+PFBS/kg-d treatment group. Moreover, statistically significant histopathological findings included increased incidence of hyperplasia in the medullary and papillary tubules of the kidneys of both sexes in the 600 mg/kg-d group. Hyperplasia and necrosis were also observed in the stomachs of males and females in the 600 mg/kg-d group; however, the study authors considered this effect to be the result of repeated gavage dosing (Lieder *et al.* 2009). Based on hematological effects, including reduced red blood cell counts, reduced hemoglobin concentrations, and decreased hematocrit values, in male rats that received 200 and

600 mg K+PFBS /kg-d, the NOAEL (no-observed-adverse-effect level) value was determined to be 60 mg/kg-d (Lieder *et al.* 2009).

From the same study, USEPA (2014) derived a chronic p-RfD for PFBS. First, the chronic p-RfD for K+PFBS was derived from the published lower-bounded benchmark dose ($BMDL_{10HED}$) value of 18.9 mg/kg-d (Lieder *et al.* 2009), which was derived from observations of increased incidence of kidney hyperplasia in female rats, from the sub-chronic duration study, as the point of departure or POD (Lieder *et al.* 2009). Thus, the benchmark dose level ($BMDL_{10HED}$) value of 18.9 mg/kg-d was divided by the calculated composite uncertainty factor (UF_C) of 1,000 to give a chronic p-RfD of 0.02 mg/kg-d (Lieder *et al.* 2009). The chronic p-RfD for the PFBS free acid was then calculated by dividing the molecular weights of the free acid (300.10) by that of the salt (338.19), to give a value of 0.89. Multiplying the chronic p-RfD by 0.89 gives a chronic p-RfD for the PFBS free acid of 0.0178 mg/kg-d (USEPA 2014, Lieder *et al.* 2009).

MAMMALIAN TOXICITY – OTHER

Mammalian Toxicity – Other: Mutagenicity

In a National Toxicology Program (NTP) study (NTP 2005), PFBS proved negative for mutagenic activity in *Escherichia coli* strain pKM101, and *Salmonella typhimurium* strain TA100, and equivocal in the Ames test for *Salmonella typhimurium* strain TA98 in the presence/absence of human liver fraction S9 under conditions of PFBS at doses of 0–5,000 µg/plate. Thus, there is no *in vitro* evidence of PFBS mutagenicity (NTP 2005).

The genotoxic potential of several PFASs, including PFBS, was analyzed in the human HepG2 cell-line – a human liver cell model system (Eriksen *et al.* 2010). HepG2 cells were treated with PFBS at 0.4–2 mM. Toxicity was determined by assaying

the ability of the PFASs to generate enhanced levels of intracellular reactive oxygen species (ROS) and pro-oxidative DNA strand-break damage in HepG2 cells. In a dose-dependent study design, PFBS was found not to induce either ROS production or DNA damage in the human liver cell model (Eriksen *et al.* 2010).

Mammalian Toxicity – Other: Endocrine Effects

Multi-generational studies of the potential disruption of the thyroid endocrine system by PFBS were carried out on the marine fish, medaka (*Oryzias melastigma*), wherein significant disruption of the thyroidal axis of the F0 marine medaka generation and disruption of thyroid function of subsequent generations was noted (Chen *et al.* 2018). On the other hand, no studies were reported that explored endocrine-disrupting effects in mammals in response to exposure to the PFBS free acid.

MAMMALIAN INHALATION TOXICITY

No published studies were identified that explored the inhalational toxicity of PFBS as the free acid.

MAMMALIAN DERMAL TOXICITY

No studies have been published which investigated the dermal toxicity of PFBS as the free acid.

AVIAN PFBS TOXICITY

In an attempt to identify the potential hazards of PFBS to avian wildlife, acute dietary studies with juvenile mallard and northern bobwhite were conducted, and these studies were complemented by a chronic dietary study of the effect of PFBS on reproduction in bobwhites (Newsted *et al.* 2008). For the acute toxicology studies, 10-day-old mallard and bobwhites were exposed to dietary concentrations of 0, 1,000, 1,780, 3,160, 5,620, or 10,000 mg PFBS (free acid)/kg wet weight (ww) feed for five days, following which the birds were fed a normal, untreated diet and observed for up to 17 days.

No treatment-related mortalities in either mallards or northern bobwhites were observed in this study at PFBS concentrations of up to 10,000 mg PBSF/kg ww feed (Newsted *et al.* 2008). Body weight gains of bobwhites exposed to 5,620 or 10,000 mg PFBS/kg feed were statistically lower than those achieved by their control counterparts. Weight gain of mallard chicks exposed to 10,000 mg PFBS/kg feed was also lower than that of their control counterparts (Newsted *et al.* 2008). For acute toxicological studies in mallard and bobwhite, the NOAEL values being 5,620 and 3,160 mg PFBS/kg ww feed, respectively.

In the reproduction component of the same study, adult bobwhites were exposed to dietary concentrations of 100, 300, or 900 mg PFBS/kg ww feed for up to 21 weeks (Newsted *et al.* 2008). No treatment-related mortalities or effects on body weight, weight gain, feed consumption, histopathological assessments, or reproductive parameters were observed in this study when compared to their control counterparts. The concentration of PFBS in blood serum, liver, and eggs was dose dependent, but less than the amount administered in the feed; moreover, liver PFBS concentrations in both adult male and female quail were approximately 30-fold lower than those in the feed. Additionally, no significant differences were found between male and female birds, suggesting that the reproductive condition of the birds might not be an important factor when predicting bird tissue PFBS concentrations. Collectively, these observations were suggestive of biodiminution. The reproductive study carried out in quail permitted the derivation of a dietary no observed adverse effect concentration of 900 mg PFBS/kg ww feed, which was stated to be equivalent to an average daily intake (ADI) of 87.8 mg PBSF/kg body weight per day (Newsted *et al.* 2008).

Additionally, biomagnification of PFBS into avian species in response to fish consumption was determined from dietary reproduction studies conducted on bobwhite (Newsted *et al.* 2008, Martin *et al.* 2003b, Martin *et al.* 2003c). These laboratory studies indicated that PFBS did not bioaccumulate following dietary exposures (Martin *et al.* 2003c). PFBS was thus deemed not to biomagnify into upper trophic-level fish or birds, wherein the bioconcentration factor (BCF) was determined to be ≤ 1.0 and the bioaccumulation factor (BAF) was assumed to be 1.0, respectfully (Martin *et al.* 2003b).

AMPHIBIAN PFBS TOXICITY

No studies have been reported that investigated acute or chronic toxicity of the PFBS free acid in amphibians.

REPTILIAN PFBS TOXICITY

In the literature review, no papers describing studies on acute or chronic toxicity of the PFBS free acid in reptiles were recorded.

RECOMMENDED PFBS TOXICITY REFERENCE VALUES

PFBS TOXICITY REFERENCE VALUES FOR MAMMALS

PFBS Toxicity Reference Values for Mammals – Oral

Comparatively few acute studies have been conducted on the effects of PFBS as the free acid in mammals, whereas no studies were identified from the primary literature on the sub-chronic or chronic effects of PFBS free acid.

The acute oral PFBS LD_{50} in male rats was determined to be 430 mg/kg (Table 7.2; Bomhard and Loser 1996). Acute effects were observed after 30 minutes post

TABLE 7.2

Summary of Oral Toxicity of PFBS (Free Acid) in Birds

| | | | Test Results | | |
| | | | | | |
Test Organism	LD_{50} (mg/kg)	NOAEL (mg/kg)	LOAEL (mg/kg)	Effects Observed at the LOAEL	Reference
Juvenile mallard and northern bobwhites	ND	5,620 (mallard) 3,160 (bobwhites)	ND	Acute toxicity component to the study. Body weight gains of exposed bobwhite were statistically lower than the controls. Weight gain of exposed mallard was lower than their control counterparts	Newsted *et al.* 2008
Male and female northern bobwhites	ND	900	ND	Reproductive component to the study. No significant effects observed	Newsted *et al.* 2008

ND – not determined

TABLE 7.3
Selected Ingestion TRVs for Mammalian Species

*TRV	Dose (mg/kg-d)	Confidence
*TRV-$_{LOW}$	0.5	Very Low
*TRV-$_{HIGH}$	1.0	Very Low

TRVs determined by the NOAEL/LOAEL approach.

exposure and by day 14, some animals showed persistent evidence of lung effects (Bomhard and Loser 1996). NOAEL values were not reported for these datasets.

Only the acute oral PFBS toxicity data, with an LD$_{50}$ value of 430 mg/kg, was marginally useful from this study for derivation of a toxicity reference value (TRV). The technical guide TG254 (USACHPPM 2000) provides an uncertainty factor of 100 for the NOAEL based on the LD$_{50}$ to give a NOAEL TRV of 4.3 mg/kg-d and a LOAEL (lowest-observed-adverse-effect level) TRV of 430 mg/kg-d (Table 7.3).

PFBS Toxicity Reference Values for Mammals – Dermal

No TRV values for dermal exposure in mammals have been reported.

PFBS TOXICITY REFERENCE VALUES FOR AVIAN SPECIES

The reproductive studies conducted in adult northern bobwhite showed that no treatment-related mortalities or effects on body weight, weight gain, feed consumption, histopathological assessments, or reproductive parameters were observed, when compared with the corresponding control birds (Newsted et al. 2008). The reproductive study that was conducted on these bobwhites permitted the derivation of a dietary NOAEL of 900 mg PFBS/kg ww feed, which was the maximal dose tested in this dose-dependent study. Thus, a TRV was not determined from this part of the study. However, for the acute toxicological studies that were conducted on juvenile mallard and bobwhites, the NOAEL values were calculated as 5,620 and 3,160 mg PFBS/kg ww feed, respectively (Newsted et al. 2008). Thus, the acute exposure NOAEL of 3,160 mg/kg that was determined for northern bobwhite was adopted for development of TRV (Table 7.4).

TABLE 7.4
Selected Ingestion TRVs for Avian Species

*TRV	Dose (mg/kg-d)	Confidence
*TRV-$_{LOW}$	105	Low
**TRV-$_{HIGH}$	3,160	Low

* TRVs determined by the NOAEL/LOAEL approach.

Additionally, the derivation of the NOAEL accounted for changes in body weight in the juvenile birds from a defined dose-response study design (Newsted *et al.* 2008). Since the chronic reproductive study showed no discernible effects on juvenile mallard or bobwhites, the acute effects obtained from studies on bobwhites were used to derive a TRV approximation. TG254 (USACHPPM, 2000) provides an uncertainty factor of 30 for the NOAEL-based TRV, but no uncertainty factor for the LOAEL TRV, since the TRVs are approximated from an acute NOAEL. Use of an uncertainty factor of 30 and the NOAEL of 3,160 mg/kg provides a NOAEL TRV of 105.3 mg/kg and a LOAEL TRV of 3,160 mg/kg. Since this TRV is derived from only one acute exposure study, and is an approximate derivation, confidence in the TRVs is low.

PFBS TOXICITY REFERENCE VALUES FOR AMPHIBIANS

No published TRV values for amphibians have been reported.

PFBS TOXICITY REFERENCE VALUES FOR REPTILES

No papers reporting TRV values for amphibians were identified in the literature search.

FUTURE RESEARCH DIRECTIONS

The toxicology of PFBS as the free acid is characterized by a marked paucity in the available data, unlike the situation with respect to its potassium or ammonium salts, for which there is a rich dataset for the acute, sub-chronic, and chronic toxicological end-points across species, covering the major classes of animals of relevance to this wildlife toxicity assessment (WTA). Indeed, with the exception of TRV derivations for the class Aves, the relative lack of available data and the poor quality of whatever data are available, which would otherwise have been used to calculate well-defined LD_{50} values and bounded NOAEL/LOAEL (or the alternative benchmark dose equivalents) derivations, precluded the generation of mammalian TRVs for this assessment report.

We found no specific data on the toxicological effects on either amphibians or reptiles of PFBS as the free acid. In addition, even though PFBS is environmentally persistent, the data suggest that it bioaccumulates or biomagnifies only weakly in terrestrial organisms. Thus, there is heightened confidence that PFBS is unlikely to be an environmental toxicant of significant concern, given the available data and the negative outcomes of the mutagenicity/genotoxicity studies reported above.

Hence, more toxicological studies of the compound are recommended – particularly, repeat-dose chronic exposure studies in mammals and birds. In addition, no published studies have investigated the effects of PFBS on wild mammal or avian species, at least up to the final literature review cross-check for this chapter, conducted on April 2, 2019. No studies were found which investigated the impacts of PFBS exposure on amphibians or reptiles, and this represents a significant data gap in developing TRVs for PFBS as the free acid. Moreover, herpetofauna are very likely to be impacted by PFBS due to its ease of waterborne transport (i.e., *via* adsorption) in the environment. Thus, future studies that focus on acute, sub-chronic, and chronic toxicity studies on wild mammalian species, as well as non-mammalian wildlife, such as birds, reptiles, and amphibians, are thoroughly warranted and recommended.

8 6:2 Fluorotelomer Sulfonate (6:2 FTS)

Allison M. Narizzano and Michael J. Quinn Jr.

CONTENTS

Compared with perfluorooctanesulfonic acid (PFOS), 6:2 fluorotelomer sulphonate (6:2 FTS) is less toxic in both rodent and other animal models, is less environmentally persistent, and has not been shown to bioaccumulate in species tested to date. Toxicology studies indicate that 6:2 FTS does not cause damage to DNA, does not act as a skin sensitizer, and does not cause toxicity to reproductive organs or to the developing fetus. Several acute toxicity studies (i.e., single-dose studies testing for lethality) have demonstrated that 6:2 FTS is less acutely toxic than PFOS in laboratory animals. It has been shown to cause skin irritation; however, this requires dermal contact at high concentrations. Rodent studies have demonstrated that exposure to 6:2 FTS can cause kidney and liver toxicity; however, these results occur at higher exposure levels than would typically be seen in environmental settings. Laboratory studies of longer duration, multiple exposure levels, and exposure studies with additional end points are still needed.

Unlike some per- and polyfluoroalkyl substances (PFAS), such as PFOS and perfluorooctanoic acid (PFOA), which are highly persistent and do not degrade in the environment, 6:2 FTS can be degraded by bacteria in the environment under certain conditions. Studies have consistently demonstrated that degradation of 6:2 FTS leads to the formation of over a dozen intermediate PFAS, including the short-chain perfluorocarboxylic acids (PFCAs) perfluorohexanoic acid (PFHxA), and perfluoropentanoic acid (PFPeA), during the degradation process(Wang et al., 2011; (Harding et al., 2015). Although PFHxA has been shown to pose minimal risk to human health, very little information on the toxicity of the other PFAS formed during 6:2 FTS degradation is available, and the effects of these PFAS on ecological and human health need to be better understood.

MAMMALIAN TOXICITY

A study summarized by ECHA reported the acute oral median lethal dose (LD_{50}) of 6:2 FTS in female rats to be between 300 and 2,000 mg/kg (ECHA 2019a). An *ex vivo* skin corrosion test suggested that 6:2 FTS is corrosive to skin. In a local lymph node assay, 6:2 FTS did not produce a dermal sensitization response in mice.

Sheng *et al.* (2017) exposed male mice to 0 or 5 mg/kg-d 6:2 FTS for 28 days *via* oral gavage. The concentrations of 6:2 FTS in serum and liver samples of dosed animals were three orders of magnitude higher than the controls, suggesting the bioaccumulation potential and slow elimination of 6:2 FTS (Sheng *et al.* 2017). Liver weights and the activities of some liver enzymes were increased in 6:2 FTS-exposed animals. Histopathological changes (i.e., hepatocellular hypertrophy and necrosis) suggest early liver damage in response to 6:2 FTS exposure. Lack of PPARα gene expression (measured by mRNA expression) implies non-activation of this pathway whereas PPARγ and related proteins were up-regulated in treated animals, compared to controls, and may be responsible for the observed liver effects.

A 14-day repeated-dose oral gavage study was conducted in male and female rats exposed to 0, 10, 50, or 100 mg/kg-d 6:2 FTS. No treatment-related mortality occurred, although reduced food consumption and body weights occurred in males exposed to 50 and 100 mg/kg-d and in females exposed to 100 mg/kg-d (ECHA 2019a). In these same groups, creatinine and urea levels, and kidney weights increased in response to treatment.

Only one study has evaluated the effects of 6:2 FTS on reproduction and development. In a combined 90-day repeated-dose oral gavage study and reproductive/developmental screening study, male and female rats were administered 0, 5, 15, or 45 mg/kg-d 6:2 FTS. No treatment-related mortality occurred and mean body weight was comparable in all groups throughout the study, including during gestation and lactation (ECHA 2019a). Similarly, food consumption was comparable across the groups. As with the 14-day study, kidney weights and creatinine and urea concentrations were affected by exposure to 6:2 FTS. Estrous cycling, mating indices, and fertility indices were unaffected by 6:2 FTS, suggesting that 6:2 FTS does not affect fertility. Similarly, the number of live-born pups, percentage pup survival, pup weight and weight gain, anogenital distance, and nipple retention were unaffected by treatment with 6:2 FTS, suggesting it is not a developmental toxicant.

Preliminary data from a study with white-footed mice (*Peromyscus leucopus*) exposed to 6:2 FTS suggest a reduced immune response, compared to the controls (Bohannon, et al., unpublished data).

AMPHIBIAN TOXICITY

Juvenile American toads (*Anaxyrus americanus*), eastern tiger salamanders (*Ambystoma tigrinum*), and northern leopard frogs (*Lithobates pipiens*) were exposed to 6:2 FTS at dermal exposures of 0, 80, 800, or 8,000 ppb on sphagnum moss substrate (dry weight basis) (Abercrombie *et al.* 2019). Snout–vent length was reduced in all three species. Scaled mass index was higher than in the controls in only northern leopard frogs. No no-observable-adverse-effect concentrations were identified for these effects, and, although the authors did not identify lowest-observable-adverse-effect concentrations for each end point, they noted that they were generally at sphagnum moss concentrations of between 50 and 120 ppb. Bioaccumulation of 6:2 FTS was considered to be minimal in this test design. Initial concentrations of 6:2 FTS in the substrate decreased significantly in all three species compared with controls, and the authors suggest that the amphibians were not only exposed to reduced levels of the PFAS over time, but to a complex mixture of PFAS, which was produced by bacterial degradation.

9 Perfluorodecanoic Acid (PFDA)

Michael J. Quinn Jr.

CONTENTS

Perfluorodecanoic acid (PFDA, or sometimes PFDeA or Ndfda) is classified as a perfluorinated carboxylic acid and is used in commercial wetting agents and aqueous film forming foams (AFFF). It has been found in environmental samples contaminated by AFFF use, and in human maternal and cord serum samples collected at delivery (Yang *et al.* 2016). The half-life of PFDA in laboratory animals and humans is unknown. There is very little toxicity information published for PFDA, although a few publications are valuable in the assessment of its potential to affect development and endocrine function (Harris and Birnbaum 1989, Van Rafelghem et al. (1987). Toxicity information of PFDA is limited to the class Mammalia.

MAMMALIAN TOXICITY

Oral median lethal doses (LD_{50})values for PFDA have been determined as 57 mg/kg for rats (George and Anderson 1986) and 111 mg/kg for mice (Harris *et al.* 1989).

To determine the potential for developmental effects of PFDA, C57BL6N mice were exposed to PFDA by gavage on gestation days (GD) 10–13 at levels of 0, 0.25, 0.5, 1.0, 2.0, 4.0, 8.0, 16.0, or 32.0 mg/kg-d or on GD 6–15 at levels of 0, 0.03, 0.3, 1.0, 3.0, 6.4, or 12.8 mg/kg-d (Harris and Birnbaum 1989). Dams were euthanized on GD 18. PFDA caused a reduction in maternal body weight gain at 6.4 and 12.8 mg/kg-d (GD 6–15) and at 16.0 and 32.0 mg/kg-d (GD 10–13). Fetal viability was reduced only in groups with dams exhibiting significant loss of body weight. Fetal body weights were also reduced at 0.1 mg/kg-d (GD 6–15) and 0.5 mg/kg-d (GD 10–13). No other developmental abnormalities were observed and the only developmental effects observed in fetuses were seen at treatment levels that were toxic to the dams.

Female Harlan Sprague Dawley rats were exposed to PFDA at 0–2.0 mg/kg by oral gavage daily for 28 days (Frawley *et al.* 2018). Female B6C3F1/N mice were exposed once per week to 0–5.0 mg/kg by oral gavage for 28 days. PFDA caused hepatocyte necrosis and hepatomegaly in rats exposed to 0.5 mg/kg-d. In mice, hepatomegaly was observed at exposure levels ≥0.625 mg/kg-wk. Decreases in splenic mass, total spleen cells, and immunoglobulin 'G' positive (Ig +) and natural killer positive (NK +) cells were observed at 5.0 mg/kg-wk. Splenic CD3+, CD4+, CD8+, and Mac3+ cell numbers decreased at ≥ 1.25 mg/kg-wk. Macrophage phagocytosis decreased in the liver of rats exposed to treatment levels ≥0.25 mg/kg-d.

A single intraperitoneal (IP) injection of PFDA to male rats caused reductions of serum thyroxine (T4) and triiodothyronine (T3) at 12 hours after PFDA treatment, and reductions in feed intake, body weight, resting heart rate, and body temperature remained at that level throughout the 8-day study. Peterson (1990) argues, however, that, despite hypothyroxinemia, the PFDA-treated rats were not functionally hypothyroid, and that any changes observed in functional thyroid status can be associated with the overt toxicity of PFDA (i.e., severe hypophagia and body weight loss). Peterson (1990) cited a study by Van Rafelghem *et al.* (1987), where rats were exposed to PFDA via IP injections at 20, 40, or 80 mg/kg. Seven days after dosing, plasma T4 concentration and free thyroxine index were markedly reduced at all doses. Plasma T3 concentrations were not affected by PFDA treatment, although pair-feeding at the high-dose level (80 mg PFDA/kg) revealed a significant reduction in T3 concentration. PFDA treatment resulted in a small decrease in basal metabolic rate, but a greater reduction in basal metabolic rate was observed in vehicle-treated controls pair-fed with rats of the 80 mg PFDA/kg dose group. Thermogenesis, as measured by oxygen consumption and body core temperatures, was not greatly affected by PFDA treatment, and these changes were paralleled by pair-feeding. Van Rafelghem *et al.* (1987) maintain that, despite alterations observed in thyroid hormone levels, the pattern of effects of PFDA exposure on functional thyroid status are inconsistent.

CONCLUSIONS

There is a lack of data on the toxicity of most PFAS to wildlife species; therefore, more toxicological studies are needed, particularly for under-represented species, such as birds, reptiles, and amphibians. There are also many published studies that prevent the development of toxicity reference values (TRVs) or lend themselves to TRVs of low confidence. For example, the many acute toxicity studies preclude the opportunity to establish no-observed-adverse-effect levels (NOAELs) by employing only a single dose. Other studies employ multiple doses, yet still use NOAELs or lowest-observed-adverse-effect levels (LOAELs) rather than the benchmark dose approach for describing the threshold for adverse effects, which is inferior for deriving TRVs.

Inhalation and dermal toxicity testing data is limited for laboratory mammals and is almost completely lacking for all other animal classes. Severely limited data are available for amphibians and reptiles, and the limited avian studies are mostly restricted to egg-injection studies. Studies that focus on both acute and chronic toxicity studies on wild mammals, as well as on non-mammalian wildlife, such as birds, reptiles, and amphibians, are particularly warranted. Further studies are needed to identify the novel pathways regulating the potential for endocrine-disrupting effects of PFAS and the doses at which they are active and relevant. Similarly, recent insights (described in other chapters) into the immunotoxicity of some PFAS and their potential to suppress host immunity and increase susceptibility to infectious diseases require additional research to determine mechanisms of action and species at risk. Finally, the lower limits of exposure that will lead to chronic adverse effects remain poorly understood.

References

3M. 2000a. Phase-out plan for POSF-based products. Docket AR226-0588. St Paul, MN: U.S. Environmental Protection Agency, USEPA.

3M. 2000b. Soil adsorption/desorption study of potassium perfluorooctane sulfonate (PFOS). Docket AR 226-1030a 1030. St Paul, MN: US Environmental Protection Agency, USEPA.

Abbott, B. D., Wolf, C. J., Schmid, J. E., et al. 2007. Perfluorooctanoic acid-induced developmental toxicity in the mouse is dependent on expression of peroxisome proliferator-activated receptor-alpha. *Toxicological Sciences* 98(2):571–581.

Abbott, B. D., Wood, C. R., Watkins, A. M., et al. 2012. Effects of perfluorooctanoic acid (PFOA) on expression of peroxisome proliferator-activated receptors (PPAR) and nuclear receptor-regulated genes in fetal and postnatal CD-1 mouse tissues. *Reproductive Toxicology* 33:491–505.

Abdellatif, A. G., Preat, V., Vamecq, J., Nilsson, R., and M. Roberfroid. 1990. Peroxisome proliferation and modulation of rat liver carcinogenesis by 2,4-dichlorophenoxyacetic acid, 2,4,5-trichlorophenoxyacetic acid, perfluorooctanoic acid and nafenopin. *Carcinogenesis* 11(11):1899–1902.

Abdellatif, A. G., Preat, V., Taber, H. S., and M. Roberfroid. 1991. The modulation of rat liver carcinogenesis by perfluorooctanoic acid, a peroxisome proliferator. *Toxicology and Applied Pharmacology* 111:530–537.

Abercrombie, S. A., de Perre, C., Choi, Y. J., et al. 2019. Larval amphibians rapidly bioaccumulate poly- and perfluoroalkyl substances. *Ecotoxicology and Environmental Safety* 178:137–145. doi: 10.1016/j.ecoenv.2019.04.022

Agency for Toxic Substances and Disease Registry (ATSDR). 2009. *Draft toxicological profile for perfluoroalkyls.* Atlanta, GA: US Department of Health and Human Services, Public Health Service, pp. 404.

Ahmed, D. Y., and M. R. Abd Ellah. 2012. Effect of exposure to perfluorooctanoic acid on hepatic antioxidants in mice. *Comparative Clinical Pathology* 21:1643–1645.

Ahrens, L., Barber, J. L., Xie, Z., and R. Ebinghaus. 2009a. Longitudinal and latitudinal distribution of perfluoroalkyl compounds in the surface water of the Atlantic Ocean. *Environmental Science and Technology* 43:3122–3127.

Ahrens, L., Yamashita, N., Yeung, L. W. Y., et al. 2009b. Partitioning behaviour of per- and polyfluoroalkyl compounds between pore water and sediment in two sediment cores from Tokyo Bay, Japan. *Environmental Science and Technology* 43:6969–6975.

Ahrens, L., Taniyasu, S., Yeung, L. W. Y., et al. 2010. Distribution of polyfluoroalkyl compounds in water, suspended particulate matter and sediment from Tokyo Bay, Japan. *Chemosphere* 79:266–272.

Ahrens L. 2011. Polyfluoroalkyl compounds in the aquatic environment: A review of their occurrence and fate. *Journal of Environmental Monitoring* 13:20–31.

Albrecht, P. P., Torsell, N. E., Krishnan, P., et al. 2013. A species difference in the peroxisome proliferator-activated receptor α-dependent response to the developmental effects of perfluorooctanoic acid. *Toxicological Sciences* 131(2):568–582.

Ali, N., Ali, L., Mehdi, T., et al. 2013. Levels and profiles of organochlorines and flame retardants in car and house dust from Kuwait and Pakistan: Implication for human exposure via dust ingestion. *Environment International* 55:62–70.

Andersen, M. E., Clewell, H. J., Tan, Y. M., et al. 2006. Pharmacokinetic modeling of satu-
rable, renal resorption of perfluoroalkylacids in monkeys—probing the determinants of
long plasma half-lives. *Toxicology* 227:156–164.

Andersen, M. E., Butenhoff, J. J., Chang, S.-C., et al. 2008. Perfluoroalkyl acids and related
chemistries—toxicokinetics and modes of action. *Toxicological Sciences* 102(1):3–14.

Anderson, R. H., Long, G. C., Porter, R. C., and J. K. Anderson. 2016. Occurrence of select
perfluoroalkyl substances at U.S. Air Force aqueous film-forming foam release sites
other than fire-training areas: Field-validation of critical fate and transport properties.
Chemosphere 150:678–685. doi: 10.1016/j.chemosphere.2016.01.014

Andersson, A. M., Jorgensen, N., Frydelund-Larsen, L., Rajpert-De Meyts, E., and
Skakkebaek, N. E. 2004. Impaired Leydig cell function in infertile men: A study
of 357 idiopathic infertile men and 318 proven fertile controls. *Journal of Clinical
Endocrinology and Metabolism* 89:3161–3167.

Ankley, G. T., Kuehl, D. W., Kahl, M. D., Jensen, K. M., Butterworth, B. C., and J. W. Nichols.
2004. Partial life-cycle toxicity and bioconcentration modeling of perfluorooctane sulfonate in the northern leopard frog (*Lithobates pipiens*). *Environmental Toxicology and
Chemistry* 23:2745–2755.

Armitage, J., Cousins, I., Buck, R. C., Prevedouros, K., Russell, M. H., MacLeod, M., and S.
H. Korzeniowski. 2006. Modeling global-scale fate and transport of perfluorooctanoate
emitted from direct sources. *Environmental Science and Technology* 40:6969–6975.
doi: 10.1021/es0614870

Armitage, J. M., Schenker, U., Scheringer, M., Martin, J. W., MacLeod, M., and I. T. Cousins.
2009. Modeling the global fate and transport of perfluorooctane sulfonate (PFOS) and
precursor compounds in relation to temporal trends in wildlife exposure. *Environmental
Science and Technology* 43:9274–9280.

Asakawa, A., Toyoshima, M., Harada, K. H., Fujimiya, M., Inoue, L., and A. Koizumi.
2008. The ubiquitous environmental pollutant perfluorooctanoic acid inhibits feeding
behavior via peroxisome proliferator-activated receptor-α. *International Journal of
Molecular Medicine* 21:439–445.

ASTSWMO (Association of State and Territorial Solid Waste Management Officials). 2015.
Perfluorinated chemicals (PFCS): Perfluorooctanoic acid (PFOA) and perfluorooctane
sulfonate (PFOS) information paper. ASTSWMO, 1101 17TH Stree, NW, Suite 707,
Washington DC 20036. https://safe.menlosecurity.com/doc/docview/viewer/docN4
09B7D6A0185f164cbd76f6d81dfa15e8e111677d6eb750f9bb2c89216d3d810dacfcae
c08b0

ATSDR (Agency for Toxic Substances and Disease Registry). 2009. *Draft toxicological
profile for perfluoroalkyls*. Atlanta, GA: Division of Toxicology and Environmental
Medicine/Applied Toxicology Branch. US Department of Health and Human Services,
404.

ATSDR (Agency for Toxic Substances and Disease Registry). 2015. *Draft toxicological
profile for perfluoroalkyls*. Atlanta, GA: Division of Toxicology and Environmental
Medicine/Applied Toxicology Branch. US Department of Health and Human Services,
574.

Austin, M. E., Kasturi, B. S., Barber, M., Kannan, K., MohanKumar, P. S., and S. M.
MohanKumar. 2003. Neuroendocrine effects of perfluorooctane sulfonate in rats.
Environmental Health Perspectives 111(12):1485–1489.

Barber, J. L., Berger, U., Chaemfa, C., Huber, S., Jahnke, A., Temme, C., and K. C. Jones. 2007.
Analysis of per-and polyfluorinated alkyl substances in air samples from Northwest
Europe. *Journal of Environmental Monitoring* 9:530–541. doi: 10.1039/b701417a

Barton, C.A., Butler, L.E., Zarzecki, C.J. et al. 2006. Characterizing perfluorooctanoate in ambient air near the fence line of a manufacturing facility: Comparing modeled and monitored values. *Journal of the Air and Waste Management Association* 56 (1): 48–55.

Beach, S. A., Newsted, J. L., Coady, K., and J. P. Giesy. 2006. Ecotoxicological evaluation of perflurooctane sulfonate (PFOS). *Reviews of Environmental Contamination and Toxicology* 186:133–174.

Beesoon, S., Webster, G. M., Shoeib, M., Harner, T., Benskin, J. P., and J. W. Martin. 2011. Isomer profiles of perflurorochemicals in matched maternal, cord, and house dust samples: Manufacturing sources and transplacental transfer. *Environmental Health Perspectives* 119(11):1659–1664. doi: 10.1289/ehp.1003265

Benninghoff, A. D., Orner, G. A., Buchner, C. H., Hendricks, J. D., and D. E. Williams. 2012. Promotion of hepatocarcinogenesis by perfluoroalkyl acids in rainbow trout. *Toxicological Sciences* 125:69–78. doi: 10.1093/toxsci/kfr267

Berthiaume, J., and K. B. Wallace. 2002. Perflurooctane, perflurooctane sulfonate, and N-ethyl per-flurooctanesulfonamido ethanol, peroxisome proliferation and mitochondrial biogenesis. *Toxicology Letters* 129:23–32.

Biegel, L. B., Liu, R. C. M., Hurtt, M. E., and J. C. Cook. 1995. Effects of ammonium perfluorooctanoate on Leydig cell function: *In vitro, in vivo*, and *ex vivo* studies. *Toxicology and Applied Pharmacology* 134:18–25.

Biegel, L. B., Hurtt, M. E., Frame, S. R., O'Connor, J. C., and J. C. Cook. 2001. Mechanisms of extrahepatic tumor induction by peroxisome proliferators in male CD rats. *Toxicological Sciences* 60:44–55.

Biesemeier, J. A., and D. L. Harris. 1974. Primary skin irritation and primary eye irritation in albina rabbits, sample T-1117 (FC-95), WARF Institute Inc. USEPA AR226-0647.

Bijland, J. A., Rensen, P. C. N., Pieterman, E. J., et al. 2011. Perfluoroalkyl sulfonates cause alkyl chain length dependent heatic steatosis and hypo-lipidemia mainly by impairing lipoprotein production in APOE 3- Leiden CETP mice. *Toxicological Sciences* 123:290–303.

Björklund, J. A., Thuresson, K. and C. A. De Witt. 2009. Perfluoroalkyl compounds (PFCs) in indoor dust: Concentrations, human exposure estimates, and sources. *Environmental Science and Technology* 43:2276–2281.

Bogdanska, J., Borg, D., Sunström, M., et al. 2011. Tissue distribution of ^{35}S-labelled per-flurooctane sulfonate in adult mice after oral exposure to a low environmentally relevant dose or a high experimental dose. *Toxicology* 284:54–62.

Bogdanska, J., Sundström, M., Bergström, U., et al. 2014. Tissue distribution of ^{35}S-labelled perfluorobutanesulfonic acid in adult mice following dietary exposure for 1–5 days. *Chemosphere* 98:28–36.

Bomhard, E., and E. Loser. 1996. Acute toxicologic evaluation of perfluorobutanesulfonic acid. [Abstract]. *Journal of the American College of Toxicology* 15:1.

Bost, P. C., Strynar, M. J., Reiner, J. L., et al. 2016. U.S. domestic cats as sentinels for perfluroalkyl substances: Possible linkages with housing, obesity and disease. *Environmental Research* 151:145–153.

Braune, B. M., and R. J. Letcher. 2013. Perfluorinated sulfonate and carboxylate compounds in eggs of seabirds breeding in the Canadian Arctic: Temporal trends (1975–2011) and interspecies comparison. *Environmental Science and Technology* 47(1):616–624.

Brignole, A. J., Porch, J. R., Krueger, H. O., and R. L. Van Hoven. 2003. PFOS: A toxicity test to determine the test substance on seedling emergence of seven species of plants. Toxicity to terrestrial plants. USEPA Docket AR226-1369. Easton, MD: Wildlife International, LTD.

Buck, R. C., Franklin, J., Berger, U., et al. 2011. Perfluoroalkyl and polyfluoroalkyl substances in the environment: Terminology, classification, and origins. *Integrated Environmental Assessment and Management* 7:513–541. doi: 10.1002/ieam.258

Butenhoff, J., Costa, G., Elcombe, C., et al. 2002. Toxicity of ammonium perfluorooctanoate in male cynomolgus monkeys after oral dosing for 6 months. *Toxicological Sciences* 69:244–257.

Butenhoff, J. L., Kennedy Jr. G. L., Frame, S. R., O'Connor, J. C., and R. G. York. 2004. The reproductive toxicology of ammonium perfluorooctanoate (PFOA) in the rat. *Toxicology* 196:95–116.

Butenhoff, J. L., Olsen, G. W., and A. Pfahles-Hutchens. 2006. The applicability of biomonitoring data for perfluorooctanesulfonate to the environmental public health continuum. *Environmental Health Perspectives* 114:1776–1782.

Butenhoff, J. L., Ehresman, D. J., Chang, S.-C., Parker, G. A., and D. G. Stump. 2009. Gestational and lactational exposure to potassium perfluorooctanesulfonate (K+PFOS) in rats: Developmental neurotoxicity. *Reproductive Toxicology* 27(3–4) Special Issue: 387–399.

Butenhoff, J. L., Chang, S. C., Olsen, G. W., and P. J. Thomford. 2012. Chronic dietary toxicity and carcinogenicity study with potassium perfluorooctanesulfonate in Sprague Dawley rats. *Toxicology* 293:1–15. doi: 10.1016/j.tox.2012.01.003

Butt, C. M., Berger, U., Bossi, R., and G. T. Tomy. 2010. Levels and trends of poly- and perfluorinated compounds in the arctic environment. *Science of the Total Environment* 408:2936–965.

Calafat, A. M., Kuklenyik, Z., Reidy, J. A., et al. 2007. Serum concentrations of 11 polyfluoroalkyl compounds in the U.S. population: Data from the National Health and Nutrition Examination Survey (NHANES). *Environmental Science and Technology* 41:2237–2242.

Case, M. T., York, R. G., and M. S. Christian. 2001. Rat and rabbit oral developmental toxicology studies with two perfluoronated compounds. *International Journal of Toxicology* 20:101–109.

Celik, A., Eke, D., Ekinci, S. Y., and S. Yildirim. 2013. The protective role of curcumin on perfluorooctane sulfonate-induced genotoxicity: Single cell gel electrophoresis and micronucleus test. *Food and Chemical Toxicology* 53:249–255. doi: 10.1016/j. fct.2012.11.054

CEMN. 2008. CEMC focussing on ionizing surfactants. Canadian Environmental Modelling Network.   www.trentu.ca/academic/aminss/envmodel/cemn/NewsReports/CEMN news200804.pdf

Chan, D. B., and E. S. K. Chian. 1986. Economics of membrane treatment of wastewaters containing firefighting foam. *Environmental Progress* 5(2):104–109.

Chang, E. T., Adami, H. O., Boffetta, P., et al. 2014. A critical review of perfluorooctanoate and perfluorooctanesulfonate exposure and cancer risk in humans. *Critical Reviews in Toxicology* 44 Supplement 1:1–81. doi: 10.3109/10408444.2014.905767

Chang, S. C., Thibodeaux, J. R., Eastvold, M. L., et al. 2008. Thyroid hormone status and pituitary function in adult rats given oral doses of perfluorooctanesulfonate (PFOS). *Toxicology* 243:330–339. doi: 10.1016/j.tox.2007.10.014

Chang, S. C., Ehresman, D. J., Bjork, J. A., et al. 2009. Gestational and lactational exposure to potassium perfluorooctanesulfonate (K + PFOS) in rats: Toxico-kinetics, thyroid hormone status, and related gene expression. *Reproductive Toxicology* 27:387–399. doi: 10.1016/j.reprotox.2009.01.005

Chang, S. C., Noker, P. E., Gorman, G. S., et al. 2012. Comparative pharmacokinetics of perfluorooctanesulfonate (PFOS) in rats, mice and monkeys. *Reproductive Toxicology* 33:428–440.

Chang S. et al. 2018. Reproductive and developmental toxicity of potassium perfluorohexane-sulfonate in CD-1 mice. *Reproductive Toxicology* 78:150–168. https://doi.org/10.1016/j .reprotox.2018.04.007.

Chen, H., Chen, S., Quan, X., Zhao, Y., and H. Zhao. 2009. Sorption of perfluorooctane sulfonate (PFOS) on oil and oil-derived black carbon, influence of solution pH and [Ca2+]. *Chemosphere* 77:1406–1411.

Chen, H., Zhang, C., Yu, Y., and J. Han. 2012. Sorption of perfluorooctane sulfonate (PFOS) on marine sediments. *Marine Pollution Bulletin* 64:902–906.

Chen, H., Reinhard, M., Nguyen, V. T., and K. Y.-H. Gin. 2016. Reversible and irreversible sorption of perfluorinated compounds (PFCs) by sediments of an urban reservoir. *Chemosphere* 144:1747–1753.

Chen, L., Hu, C., Tsui, M. M. P., et al. 2018. Multigenerational disruption of the thyroid endocrine system in marine medaka after a life-cycle exposure to perfluorobutanesulfonate. *Environmental Science and Technology* 52(7):4432–4439.

Chen, N., Li, J., Li, D., Yang, Y., and D. He. 2014. *Chronic exposure to perfluorooctane sulfonate induces behavior defects and neurotoxicity through oxidative damage, in vivo and in vitro. PlOS* 9(11): e113453. doi: 10.1371/journal.pone.0113453

Chen, S., Jiao, X.-C., Gai, N., et al. 2016. Perfluorinated compounds in soil, surface water, and groundwater from rural areas in eastern China. *Environmental Pollution* 211:124–131. doi: 10.1016/j.envpol.2015.12.024

Cheng, X. and C. D. Klaassen. 2008. Perfluorocarboxylic acids induce cytochrome p450 enzymes in mouse liver through activation of PPAR-α and CAR transcription factors. *Toxicological Sciences* 106(1):29–36.

Chengelis, C.P., et al. 2009. Comparison of the toxicokinetic behavior of perfluorohexanoic acid (PFHxA) and nonafluorobutane-1-sulfonic acid (PFBS) in cynomolgus monkeys and rats. *Reproductive Toxicology* 27(3–4):400–406. https://doi.org/10.1016/j.reprotox. 2009.01.013.

Chou, S., Jones, D., Pohl, H. R., et al. 2015. *Draft toxicological profile for perfluoro-alkyls.* , Atlanta, GA: U.S. Department of Health and Human Services, Public Health Service, Agency for Toxic Substances and Disease Registry, 1–574.

Chou, W.-C., and Z. Lin. 2019. Bayesian evaluation of a physiologically based pharmacokinetic (PBPK) model for perfluorooctane sulfonate (PFOS) to characterize the interspecies uncertainty between mice, rats, monkeys, and humans: Development and performance verification. *Environment International* 129:408–422. doi: 10.1016/j. envint.2019.03.058

Conder, J. M., Hoke, R. A., de Wolf, W., Russell, M. H., and R. C. Buck. 2008. Are PFCAs bioaccumulative? A critical review and comparison with regulatory criteria and persistent lipophilic compounds. *Environmental Science and Technology* 42(4):995–1003. doi: 10.1021/es070895g

Conder, J. M., Wenning, R. J., Travers, M., and M. Blom. 2010. Overview of the environmental fate of perfluorinated compounds. Network for Industrially Contaminated Land in Europe (NICOLE) Technical Meeting. 4 November 2010.

Cook, J. C., Murray, S. M., Frame, R. S., and M. E. Hurtt. 1992. Induction of Leydig cell adenomas by ammonium perfluorooctanoate: A possible endocrine-related mechanism. *Toxicology and Applied Pharmacology* 1134:209–217.

Coperchini, F., Awwad, O., Rotondi, M., Santini, F., Imbriani, M., and L. Chiovato. 2017. Thyroid disruption by perfluorooctane sulfonate (PFOS) and perfluorooctanoate (PFOA). *Journal of Endocrinological Investigation* 40:105–121.

Corsini, E., Sangiovanni, E., Avogadro, A., et al. 2012. *In vitro* characterization of the immunotoxic potential of several perfluorinated compounds (PFCs). *Toxicology and Applied Pharmacology* 258(2):248–55.

Cui, L., Q.-f. Zhou, C.-y. Liao, J. Fu, and G.-B. Jiang. 2009. Studies on the toxicological effects of PFOA and PFOS on rats using histological observation and chemical analysis. *Archives of Environmental Contamination and Toxicology* 56:338–349.

Cui, L., Liao, C-Y., Zhou, Q. F., et al. 2010. Exceretion of PFOA and PFOS in male rats during a sub-chronic exposure. *Archives of Environmental Contamination and Toxicology* 58:205–213.

Curran, I., Hierlihy, S. L., Liston, V., et al. 2008. Altered fatty acid homeostasis and related toxicologic sequelae in rats exposed to dietary potassium perfluorooctanesulfonate (PFOS). *Journal of Toxicology and Environmental Health* 71:1526–1541. doi: 10.1080/15287390802361763

Dai, J., Li, M., Jin, Y., Norimitsu, S., Xu, M., and F. Wei. 2006. Perfluorooctanesulfonate and perfluorooctanoate in red panda and giant panda from China. *Environmental Science and Technology* 40:5647–5652. doi: 10.1021/es0609710

Darlington, R., Barth, E., and J. McKernan. 2018. The challenges of PFAS remediation. *Military Engineer* 110:58–60.

Darwin, R. L., Ottman, R. E., Norman, E. C., Gott, J. E., and Hanauska, C. P. 1995. Foam and the environment: A delicate balance. *Fire Journal* 89(3):67–73.

Darwin, R. L. 2011. *Estimated inventory of PFOS-based aqueous film forming foam (AFFF)*. Arlington, VA: Fire Fighting Foam Coalition.

Das, K. P., Grey, B. E., Rosen, M. B., et al. 2015. Developmental toxicity of perfluorononanoic acid in mice. *Reproductive Toxicology* 51:133–144.

Dauwe, T., Van de Vijver, K., De Coen, W., and M. Eens. 2007. PFOS levels in the blood and liver of a small songbird near a fluorochemical plant. *Environment International* 33:357–361.

Dean, W., Jessup, D., Thompson, G., Romig, G., and D. Powell. 1978. Fluorad fluorochemical surfactant FC-95 acute oral toxicity (LD_{50}) study in rats. *International Research and Development Corporation* 137(83):4582.

Dean, W.P., and D.C. Jessup. 1978. *Acute Oral Toxicity (LD50) Study in Rats*. International Research and Development Corporation, Study No. 137-091, May 5, 1978. U.S. Environmental Protection Agency *Administrative Record* 226-0419.

Dean, W. P., Jessup, D. C., Thompson, G., Romig, G., and D. Powel. 1978. Flurorad fluorochemical surfactant FC-95 acute oral toxicity (LD_{50}) study in rats. Study no. 137-083, International Research Development Corporation (as cited in OECD 2002, USEPA 2014). *Science and Technology* 35:3065–3070.

Deck, A. T., and M. S. Johnson. 2015. Methods for derivation of wildlife toxicity reference values for use in ecological risk assessments. In *Wildlife Toxicity Assessments for Chemical of Military Concern*, eds. Williams, M. A., Reddy, G., Quinn, M. J. Jr., and Johnson, M. S. Amsterdam: Elsevier, pp. 15–22.

D'eon, J. C., Hurley, M. D., Wallington, T. J., and S. A. Mabury. 2006. Atmospheric chemistry of *N*-methyl perfluorobutane sulfonamidoethanol, $C_4F_9SO_2N(CH_3)CH_2CH_2OH$: Kinetics and mechanism of reaction with OH. *Environmental Science and Technology* 40:1862–1868.

D'Eon, J. C., and S. A. Mabury. 2007. Production of perfluorinated carboxylic acids (PFCAs) from the biotransformation of polyfluoroalkyl phosphate surfactants (PAPS): Exploring routes of human contamination. *Environmental Science and Technology* 41(13):4799–4805.

DeSilva, A. O., Allard, C. N., Spencer, C., Webster, G. M., and M. Shoeib. 2012. Phosphorus-containing fluorinated organics: Polyfluoroalkyl phosphoric acid diesters (diPAPs), perfluorophosphonates (PFPAs), and perfluorophosphinates (PFPIAs) in residential indoor dust. *Environmental Science and Technology* 46(22):12575–12578.

DeVos, M. G., Huijbregts, M. A. J., van den Heuvel-Greve, M. J., et al. 2008. Accumulation of perfluorooctane sulfonate (PFOS) in the food chain of the Western Scheldt estuary: Comparing field measurements with kinetic modeling. *Chemosphere* 70:1766–1773. doi: 10.1016/j.chemosphere.2007.08.038

DeWitt, J. C., Copeland, C. B., Strynar, M. J., and R. W. Luebke. 2008. Perfluorooctanoic acid-induced immunomodulation in adult C57BL/6 J or C57BL/6 N female mice. *Environmental Health Perspectives* 116(5):644–650.

DeWitt, J. C., Copeland, C. B., Strynar, M. J., and R. W. Luebke. 2009. Suppression of humoral immunity by perfluorooctanoic acid is independent of elevated serum corticosterone concentration in mice. *Toxicological Sciences* 109(1):106–112.

DeWitt, J. C., Peden-Adams, M. M., Keller, J. M., and D. R. Germolec. 2012. Immunotoxicity of perfluorinated compounds: Recent developments. *Toxicologic Pathology* 40(2):300–311.

Dietert, R. R., DeWitt, J. C., Germolec, D. R., and J. T. Zelikoff. 2010. Breaking patterns of environmentally influenced disease for health risk reduction: Immune perspectives. *Environmental Health Perspectives* 118(8):1091-9-91.

Dixon, D., Reed, C. E., Moore, A. B., et al. 2012. Histopathologic changes in the uterus, cervix and vagina of immature CD-1 mice exposed to low doses of perfluoro-octanoic acid (PFOA) in a uterotrophic assay. *Reproductive Toxicology* 33:506–512.

Dong, G. H., Zhang, Y. H., Zheng, L., et al. 2009. Chronic effects of perfluorooctane sulfonate exposure on immunotoxicity in adult male C57BL/6 mice. *Archives of Toxicology* 83:805–815.

Dong, G. H., Liu, M. M., Wang, D., et al. 2011. Sub-chronic effect of perfluoro -octane sulfonate (PFOS) on the balance of type 1 and type 2 cytokine in adult C57BL6 mice. *Archives of Toxicology* 85(10):1235–1244.

Dong, G. H., Tung, K. Y., Tsai, C. H., et al. 2013. Serum polyfluoroalkyl concentrations, asthma outcomes, and immunological markers in a case-control study of Taiwanese children. *Environmental Health Perspectives* 121(4):507–513.

ECHA, 2019a. REACH Registration Dossier for CAS number 27619-97-2 (tridecafluorooctanesulphonic acid), in: Agency, E.C. (Ed.). European Chemicals Agency.

EFSA (European Food Safety Authority). 2008. Perfluorooctane sulfonate (PFOS), Perfluorooctanoic acid (PFOA) and their salts. Scientific opinion of the panel on contaminants in the food chain. *European Food Safety Authority* 653:1–131.

Ehresman, D. J., Froehlich, J. W., Olsen, G. W., Chang, S.-C., and J. L. Butenhoff. 2007. Comparison of human whole blood, plasma, and serum matrices for the determination of perfluorooctanesulfonate (PFOS), perfluorooctanoate (PFOA), and other fluorochemicals. *Environmental Research* 103:176–184. doi: 10.1016/j.envres.2006.06.008

Eldasher, L. M., Wen, X., Little, M. S., et al. 2013. Hepatic and renal *Bcrp* transporter expression in mice treated with perfluorooctanoic acid. *Toxicology* 306:108–113.

Eke, D., and A. Celik. 2019. Curucumin prevents perfluorooctane sulfonate-induced genotoxicity and DNA damage in rat peripheral blood. *Drug and Chemical Toxicology* 39(1):97–103.

Elcombe C. R., Elcombe, B. M., Foster. J. R., et al. 2012a. Hepatocellular hypertrophy and cell proliferation in Sprague-Dawley rats from dietary exposure to potassium perfluorooctanesulfonate results from increased expression of xenosensor nuclear receptors PPARα and CAR/PXR. *Toxicology* 293(1–3):16–29.

Elcombe, C. R., Elcombe, B. M., Foster, J. R., et al. 2012b. Evaluation of hepatic and thyroid responses in male Sprague-Dawley rats for up to eighty-four days following seven days of dietary exposure to potassium perfluorooctanesulfonate. *Toxicology* 293(1–3):30–40. doi: 10.1016/j.tox.2011.12.015

Emmett, E. A., Shofer, F. S., Zhang, H., et al. 2006a. Community exposure to perfluorooc-tanoate: Relationships between serum concentrations and exposure sources. *Journal of Occupational and Environmental Medicine* 48:759–770. doi: 10.1097/01.jom.0000232486.07658.74

Emmett, E. A., Zhang, H., Shofer, F. S., et al. 2006b. Community exposure to perfluorooc-tanoate: Relationships between serum levels and certain health parameters. *Journal of Occupational and Environmental Medicine* 48(8):771–779. doi: 10.1097/01.jom.0000233380.13087.37

Emmett, E. A., Zhang, H., Shofer, F. S., et al. 2009. Development and successful application of a "community- first" communication model for community-based environmental health research. *Journal of Occupational and Environmental Hygiene* 51(2):146–156. doi: 10.1097/JOM.0b013e3181965d9b

Eriksen, K. T., Raaschou-Nielsen, O., Sørensen, M., et al. 2010. Genotoxic potential of the perfluorinated chemicals PFOA, PFOS, PFBS, PFNA and PFHxA in human HepG2 cells. *Mutation Research* 700(1–2):39–43.

Eschauzier, C., Beerendonk, E., Scholte-Veenendaal, P., and P. De Voogt. 2012. Impact of treatment processes on the removal of perfluoroalkyl acids from the drinking water production chain. *Environmental Science and Technology* 46(3):1708–1715.

Fàbrega, F., Kumar, V., Benfenati, E., et al. 2015. Physiologically based pharmacokinetic modeling of perfluoroalkyl substances in the human body. *Toxicological andEnviron-mental Chemistry* 97(6):814–827.

Fair, P. A., Driscoll, E., Mollenhauer, M. A. M., et al. 2011. Effects of environmentally rele-vant levels of perflurooctane sulfonate on clinical parameters and immunological func-tions in B6C3F1 mice. *Immunotoxicology* 8(10):17–29.

Fair, P. A., Houde, M., Hulsey, T. C., et al. 2012. Assessment of perfluorinated compounds (PFCs) in plasma of bottlenose dolphins from two southeast US estuarine areas: Relationship with age, sex and geographic locations. *Marine Pollution Bulletin* 64:66–74.

Fair, P. A., Romano, T., Schaefer, A. M., et al. 2013. Associations between perfluoroal-kyl compounds and immune and clinical chemistry parameters in highly exposed bottlenose dolphins (*Tursiops truncatus*). *Environmental Toxicology and Chemistry* 32(4):736–746.

Fairley, K. J., Purdy, R., Kearns, S., Anderson, S. E., and B. J. Meade. 2007. Exposure to the immunosuppresant, perfluorooctanoic acid, enhances the murine IgE and airway hyperreactivity response to ovalbumin. *Toxicological Sciences* 97(2):375–383.

Falandysz, J., Taniyasu, S., Gulkowaska, A., Yamashita, N., and U. Schulte-Oehlmann. 2006. Is fish a major source of fluorinated surfactants and repellents in humans living on the Baltic Coast? *Environmental Science and Technology* 40:748–751.

Falandysz, J., Taniyasu, S., Yamashita, N., Zalewski, K., and K. Kannan. 2007. Perfluorinated compounds in some terrestrial and aquatic wildlife species from Poland. *Journal of Environmental Science and Health, Part A,* 42(6):715–719. doi: 10.1080/10934520701304369

Falk, S., Brunn, H., Schröter-Kermani, C., et al. 2012. Temporal and spatial trends of perfluo-roalkyl substances in liver of roe deer (*Capreolus capreolus*). *Environmental Pollution* 171:1–8.

Fang, X., Y. Feng, Z, Shi, and J. Dai. 2009. Alterations of Cytokines and MAPK signaling pathways are related to the immunotoxic effect of perfluorononanoic acid. *Tixicological Sciences.* 108(2):367–376. https://doi.org/10.1093/toxsci/kfp019.

Fang, X., Y. Feng, J. Wang, and J. Dai. 2010. Perfluorononanoic acid-induced apoptosis in rat spleen involves oxidative stress and the activation of caspase-independent death path-way. *Toxicology* 267(1–3):54–59. https://doi.org/10.1016/j.tox.2009.10.020.

Fenton, S. E., Reiner, J. L., Nakayama, S. F., et al. 2009. Analysis of PFOA in dosed CD-1 mice. Part 2: Disposition of PFOA in tissues and fluids from pregnant and lactating mice and their pups. *Reproductive Toxicology* 27:365–372.

Fernández Freire, P., Pérez Martin, J. M., Herrero, O., et al. 2008. In vitro assessment of the cytotoxic and mutagenic potential of perfluorooctanoic acid. *Toxicology in Vitro* 22:1228–1233.

Filgo, A. J., Quist, E. M., Hoenerhoff, M. J., et al. 2014. Perfluorooctanoic acid (PFOA)-induced liver lesions in two strains of mice following developmental exposures: PPARα is not required. *Toxicologic Pathology* 43(4):558–568. doi: 10.1177/0192623314558463

Filipovic, M., Woldegiorgis, A., Norström, K., et al. 2015. Historical usage of aqueous film forming foam: A case study of the widespread distribution of perfluoroalkyl acids from a military airport to groundwater, lakes, soils and fish. *Chemosphere* 129:39–45.

Fire Fighting Foam Coalition. 2014. *Fact sheet on AFFF fire fighting agents.* Arlington, VA: Fire Fighting Foam Coalition. Available at: http://www.fffc.org/images/AFFFfactsheet 14.pdf [Accessed 22 March 2017].

Fisher, J. S., MacPherson, S., Marchetti, N., and R. M. Sharpe. 2003. Human 'testicular dysgenesis syndrome': A possible model using *in-utero* exposure of the rat to dibutyl phthalate. *Human Reproduction* 18:1383–1394.

Florentin, A., Deblonde, T., Diguio, N., Hautemaniere, A., and P. Hartemann. 2011. Impacts of two perfluorinated compounds (PFOS and PFOA) on human hepatoma cells: Cytotoxicity but no genotoxicity? *International Journal of Hygiene and Environmental Health* 214:493–499. doi: 10.1016/j.ijheh.2011.05.010

Franko, J., Meade, B. J., Frasch, H. F., Barbero, A. M., and S. E. Anderson. 2012. Dermal penetration potential of perfluorooctanoic acid (PFOA) in human and mouse skin. *Journal of Toxicology and Environmental Health, Part A* 75(1):50–62.

Fraser, A. J., Webster, T. F., Watkins, D. J., et al. 2012. Polyfluorinated compounds in serum linked to indoor air in office environments. *Environmental Science and Technology* 46:1209–1215.

Fraser, A. J., Webster, T. F., Watkins, D. J., et al. 2013. Polyfluorinated compounds in dust from homes, offices, and vehicles as predictors of concentrations in office workers' serum. *Environment International* 60:128–36.

Frawley, R. P., Smith, M., Cesta, M. F., et al. 2018. Immunotoxic and hepatotoxic effects of perfluoro-n-decanoic acid (PFDA) on female Harlan Sprague-Dawley rats and B 6 C 3 F 1/N mice when administered by oral gavage for 28 days. *Journal of Immunotoxicology* 15:41–52.

Freire, P., J.M. Pérez Martin, O. Herrero, A. Peropadre, E. de la Peña, and M.J. Hazen. 2008. In vitro assessment of the cytotoxic and mutagenic potential of perfluorooctanoic acid. *Toxicology In Vitro* 22:1228–1233.

Fuentes, S., Colomina, M. T., Vicens, P., Franco-Pons, N., and J. Domingo. 2007. Concurrent exposure to perfluorooctane sulfonate and restraint stress during pregnancy in mice: Effects on post-natal development and behavior of the offspring. *Toxicological Sciences* 98:589–598.

Furdui, F., Stock, N., Ellis, D.A., et al. 2007. Spatial distribution of perfluoroalkyl contaminants in lake trout from the Great Lakes. *Environmental Science and Technology* 41:1554–1559. doi: 10.1021/es0620484

Gabriel, K. 1976a. Acute oral toxicity – Rats. Biosearch, Inc., September 16, 1976. U.S. Environmental Protection Agency Administrative Record 226-0425.

Gabriel, K. 1976b. Primary eye irritation study in rabbits. Biosearch, Inc., September 16, 1976. U.S. Environmental Protection Agency Administrative Record 226-0426.

Gabriel, K. 1976c. Primary eye irritation study in rabbits. Biosearch, Inc., March 4, 1976. U.S. Environmental Protection Agency Administrative Record 226-0422.

Gabriel, K. 1976d. Primary skin irritation study in rabbits. Biosearch, Inc. March 4, 1976. U.S. Environmental Protection Agency Administrative Record 226-0423.

Gallagher, S. P., Van Hoven, R. L., Beavers, J. B., and M. Jaber. 2003a. PFOS: A reproduction study with Northern Bobwhite. Final report. Project No. 454-108. USEPA Administrative Record AR-226-1831. Easton, MD: Wildlife International, Ltd.

Gallagher, S. P., Van Hoven, R. L., Beavers, J. B., and M. Jaber. 2003b. PFOS: A reproduction study with Mallards. Final report. Project No. 454-109. USEPA Administrative Record AR-226-1836. Easton, MD: Wildlife International, Ltd.

Gallagher, S.P., Van Hoven, R.L., Beavers, J.B., 2003c. PFOS: a pilot reproductive study with the northern bobwhite. Wildlife International, Ltd., Project No. 454-104. USEPA Administrative Record AR-226-1817.

Garry, V. F., and R. L. Nelson. 1981. *An assay of cell transformation and cytotoxicity in C3H10T½ clonal cell line for the test chemical T-2942 CoC.* Minneapolis, MN: Stone Research Labs. March 4, 1981. U.S. Environmental Protection Agency Administrative Record 226-0428.

Gebbink, W. A., Bossi, R., Rigét, F. F., et al. 2016. Observation of emerging per- and poly-fluoroalkyl substances (PFASs) in Greenland marine mammals. *Chemosphere* 144:2384–2391.

George, M. E. and M. E. Anderson. 1986. Toxic effects of nonadecafluoro-n-decanoic acid in rats. *Toxicology and Applied Pharmacology* 85:169–180.

Giesy, J. P., and K. Kannan. 2001. Global distribution of perflurooctane sulfonate in wildlife. *Environmental Science and Technology* 35(7):1339–1342.

Giesy, J. P. and K. Kannan. 2002. Perflourochemical surfactants in the environment. *Environmental Science and Technology* 36(7):146A–152A.

Glaza, S. 1995. Acute dermal toxicity study of T-6342 in rabbits. Madison, WI: Corning Hazelton, Inc. Project ID: HWI 50800374. 3M Company. St. Paul, MN: U.S. Environmental Protection Agency Administrative Record, 226-0427.

Glaza, S. M. 1997. Acute oral toxicity study of T-6669 in rats. Corning Hazleton Inc. Study No. CHW 61001760, January 10, 1997. U.S. Environmental Protection Agency Administrative Record, 226–0420.

Goeritz, I., Falk, S., Stahl, T., Schaefers, C., and C. Schlechtriem. 2013. Biomagnification and tissue distribution of perfluoroalkyl substances (PFASs) in market-size rainbow trout (*Oncorhyncus mykiss*). *Environmental Toxicology and Chemistry* 32:2078–2088.

Goldenthal, E. I., Jessup, D. C., Geil, R. G., and J. S. Mehring. 1978b. Ninety-day sub-acute rhesus monkey toxicity. Study No. 137-092. Mattawan, MI: International Research and Development Corporation. FYI-0500–1378 (as cited in OECD 2002).

Goldenthal, E. I., Jessup, D. C., Geil, R. G., and J. S. Mehring. 1979. Ninety-day sub-acute rhesus monkey toxicity. Study No. 137-087. Mattawan, MI: International Research and Development Corporation. FYI-0500–1378, 1979 (as cited in OECD 2002).

Goosey, E. and S. Harrad. 2012. Perfluoroalkyl substances in UK indoor and outdoor air: Spatial and seasonal variation, and implications for human exposure. *Environment International* 45:86–90.

Gorrochategui, E., Lacorte, S., Tauler, R., F. L. Martin. 2016. Perfluoroalkylated substance effects in *Xenopus laevis* A6 kidney epithelial cells determined by ATR-FTIR spectroscopy and chemometric analysis. *Chemical Research in Toxicology* 29(5):924–32.

Guerranti, C., Ancora, S., Bianchi, N., et al. 2013. Perfluorinated compounds in blood of *Caretta caretta* from the Mediterranean Sea. *Marine Pollution Bulletin* 73(1):98–101.

Guruge, K. S., Hikono, H., Shimada, N., et al. 2009. Effect of perfluorooctane sulfonate (PFOS) on influenza A virus-induced mortality in female B6C3F1 mice. *Journal of Toxicological Sciences* 34(6):687–91.

Han, J -S., Jang, S., Son H.-Y., et al. 2020. Subacute dermal toxicity of perfluoroalkyl carbox-ylic acids: Comparison with different carbon-chain lengths in human skin equivalents and systemic effects of perfluoroheptanoic acid in Sprague Dawley rats. *Archives of Toxicology* 94:523–539.

Hanhijärvi, H., Ylinen, M., Kojo, A., and V. M. Kosma. 1987. Elimination and toxicity of per-fluorooctanoic acid during subchronic administration in the Wistar rat. *Pharmacology and Toxicology* 61:66–68.

Hansen, K., Johnson, H. O., Eldrdge, J. S., Butenhoff, J. L., and I. A. Dick. 2002. Quantitative characterization of trace levels of PFOA and PFOS in the Tennessee River. *Environmental Science and Technology* 36:1681–1685.

Harada, K., Nakanishi, S., Saito, N., et al. 2005. Airborne perfluorooctanoate may be a sub-stantial source contamination in Kyoto area, *Japan. Bull Environ Contam Toxicol* 74:64–69.

Harada, K., Nakanishi, S., Sasaki, K., et al. 2006. Particle size distribution and respiratory deposition estimates of airborne perfluorooctanoate and perfluorooctanesulfonate in Kyoto area, Japan. *Bull Environ Contam Toxicol* 76(2):306–310.

Harding-Marjanovic, K.C., E.F. Houtz, S. Yi, J.A. Field, D.L. Sedlak, and L. Alvarez-Cohen. 2015. Aerobic Biotransformation of Fluorotelomer Thioether Amido Sulfonate (Lodyne) in AFFF-Amended Microcosms. *Environ Sci Technol* 49 (13):7666–7.

Harris, M. W., Uraih, L. C., and L. S. Birnbaum. 1989. Acute toxicity of perfluorodecanoic acid in C57BL/6 mice differs from 2,3,7,8-tetrachlorodibenzo-p-dioxin. *Fundamental and Applied Toxicology* 13:723–736.

Hazardous Substances Databank (HSDB). 2016. *Perfluorobutanesulfonic acid. On-line Database.* Washington, DC: National Library of Medicine. Available at: http://toxnet .nlm.nih.gov [Accessed 05 February 2019].

Hekster, F. M., Laane, R. W. P. M., and P. de Voogt, P. 2002. Environmental and toxicity effects of perfluoroalkylated substances. *Reviews of Environmental Contamination and Toxicology* 179:99–121.

Hellsing, M. S., Josefsson, S., Hughes, A. V., and L. Ahrens. 2016. Sorption of perfluoroalkyl substances to two types of minerals. *Chemosphere* 159:385–391.

Henry, B. J., Carlin, J. P., Hammerschmidt, J. A., et al. 2018. A critical review of the applica-tion of polymer of low concern and regulatory criteria to fluoropolymers. *Integrated Environmental Assessment and Management* 14(3): 316–344. doi: 10.1002/ieam.4035

Henry, N. D., and P. A. Fair. 2013. Comparison of in vitro cytotoxicity, estrogenicity and anti-estrogenicity of triclosan, perfluorooctane sulfonate and perfluorooctanoic acid. *Journal of Applied Toxicology* 33:265–272.

Higgins, C. P., and R. G. Luthy. 2006. Sorption of perfluorinated surfactants on sediments. *Environmental Science and Technology* 40:7251–7256.

Higgins, C. P., and R. G. Luthy. 2007. Modeling sorption of anionic surfactants onto sediment materials, an a priori approach for perfluoroalkyl surfactants and linear alkylbenzene sulfonates. *Environmental Science and Technology* 41:3254–3261.

Hinderliter, P. M., Mylchreest, E., Gannon, S. A., Butenhoff, J. L., and G. L. Kennedy Jr. 2005. Perfluorooctanoate: Placental and lactational transport pharmacokinetics in rats. *Toxicology* 211:139–148.

Hinderliter, P. M., DeLorme, M. P., and G. L. Kennedy. 2006. Perfluorooctanoic acid: Relationship between repeated inhalation exposures and plasma PFOA concentration in the rat. *Toxicology* 222:80–85.

Hines, E. P., White, S. S., Stanko, J. P., et al. 2009. Phenotypic dichotomy following devel-opmental exposure to perfluorooctanoic acid (PFOA) in female CD-1 mice: Low doses induce elevated serum leptin and insulin, and overweight in mid-life. *Molecular and Cellular Endocrinology* 304:97–105.

Holm, M., Rajpert-De Meyts, E., Andersson, A. M., and N. E. Skakkebaek. 2003. Leydig cell micronodules are a common finding in testicular biopsies from men with impaired spermatogenesis and are associated with decreased testosterone/LH ratio. *Journal of Pathology* 199:378–386.

Houde, M., Wells, R. S., Fair, P. A., et al. 2005. Polyfluoroalkyl compounds in free-ranging bottlenose dolphins (*Tursiops truncatus*) from the Gulf of Mexico and the Atlantic Ocean. *Environmental Science and Technology* 39:6591–6598. doi: 10.1021/es0506556

Houde, M., Bujas, T. D., Small, J., et al. 2006a. Biomagnification of perfluoroalkyl compounds in the bottlenose dolphin (*Tursiops truncatus*) food web. *Environmental Science and Technology* 40:4138–4144.

Houde, M., Martin, J. W., Letcher, R. J., Solomon, K. R., and D. C. Muir. 2006b. Biological monitoring of polyfluoroalkyl substances: A review. *Environmental Science and Technology* 40:3463–473.

Houtz, E. F., Sutton, R., Park, J. S., and M. Sedlak. 2016. Poly- and perfluoroalkyl substances in waste water: Significance of unknown precursors, manufacturing shifts, and likely AFFF impacts. *Water Research* 95:142–149.

Hu, Q., Strynar, M., and J. DeWitt. 2010. Are developmentally exposed C57Bl/6 mice insensitive to suppression of TDAR by PFOA? *Journal of Immunotoxicology* 7(4): 344–349.

Hu, W., Jones, P. D., Upham, B. L., Trosko, J. E., Lau, C., and J. P. Geisy. 2002. Inhibition of gap junctional intercellular communication by perfluorinated compounds in rat liver and dolphin kidney epithelial cell lines in vitro and Sprague-Dawley rats *in vivo*. *Toxicological Sciences* 68(2):429–436.

Huang, M. C., Dzierlenga, A. L., Robinson, V. G., et al. 2019. Toxicokinetics of perfluorobutane sulfonate (PFBS), perfluorohexane-1-sulphonic acid (PFHxS), and perfluorooctane sulfonic acid (PFOS) in male and female Hsd: Sprague Dawley SD rats after intravenous and gavage administration. *Toxicology Reports* 6:645–655.

Hundley, S., Sarrif, A., and G. Kennedy. 2006. Absorption, distribution and excretion of ammonium perfluorooctanoate (PFOA) after oral administration in various species. *Drug and Chemical Toxicology* 29:137–145.

ICH (International Confereence on Harmonization of Technical Requirements for Registration of Pharmaceuticals for Human Use). 2005. Immunotoxicity studies for human pharmaceuticals S8 11:2005. Available at: http://www.ich.org/products/guidelines/safety/arti cle/safety-guidelines.html.

Interstate Technology Regulatory Council (ITRC). 2018. Naming conventions and physical and chemical properties of per- and polyfluoroalkyl substances (PFAS). Retrieved January 2020 from: https://pfas-1.itrcweb.org/wp-content/uploads/2018/03/pfas_fact_s heet_naming_ conventions__3_16_18.pdf.

Jacquet, N., Maire, M. A., Landkocz, Y., and P. Vasseur. 2012. Carcinogenic potency of perfluorooctane sulfonate (PFOS) on Syrian hamster embryo (SHE) cells. *Archives of Toxicology* 86:305–314. doi: 10.1007/s00204-011-0752-8

Jacquet, N., Maire, M. A., Rast, C., Bonnard, M., and P. Vasseur. 2012. Perfluorooctanoic acid (PFOA) acts as a tumor promoter on Syrian hamster embryo (SHE) cells. *Environmental Science and Pollution Research* 19:2537–2549.

Jaspers, V. L., Herzke, D., Eulaers, I., Gillespie, B. W., and M. Eens. 2013. Perfluoroalkyl substances in soft tissues and tail feathers of Belgian barn owls (*Tyto alba*) using statistical methods for left-censored data to handle non-detects. *Environment International* 52:9–16.

Jiang, Q., Lust, R. M., Strynar, M. J., Dagnino, S., and J. C. DeWitt. 2012. Perfluorooctanoic acid induces developmental cardiotoxicity in chicken embryos and hatchlings. *Toxicology* 293:97–106.

Jiang, Q., Wang, C., Xue, C., et al. 2016. Changes in the levels of L-carnitine, acetyl-L-carnitine and propionyl-L-carnitine are involved in perfluoro-octanoic acid induced developmental cardiotoxicity in chicken embryo. *Environmental Toxicology and Pharmacology* 48:116–124.

Jin, Y.H., Liu, W., Sato, I., et al. 2009. PFOS and PFOA in environmental and tap water in China. *Chemosphere* 77:605–611.

Joensen, U. N., Bossi, R., Leffers, H., et al. 2009. Do perfluoroalkyl compounds impair human semen quality? *Environmental Health Perspectives* 117:923–927.

Johansson, N., Fredriksson, A., and P. Eriksson. 2008. Neonatal exposure to perfluorooctane sulfonate (PFOS) and perfluorooctanoic acid (PFOA) causes neurobehavioural defects in adult mice. *NeuroToxicology* 29:160–169.

Johansson, N., Eriksson, P., and H. Viberg. 2009. Neonatal exposure to PFOS and PFOA in mice results in changes in proteins which are important for neuronal growth and synaptogenesis in the developing brain. *Toxicological Sciences* 108(2):412–418.

Johnson, J. D., Gibson, S. J., and R. E. Ober. 1979. Extent and route of excretion and tissue distribution of total carbon-14 in rats after a single intravenous dose of FC-95-14C. St Paul, MN: Riker Laboratories, Inc. U.S. E.P.A. public docket, administrative record AR-226-006.

Johnson, M. S., and M. J. McAtee. 2015. Characterizing potential for toxicity: Estimating risks to wildlife. In *Wildlife Toxicity Assessments for Chemical of Military Concern*, eds. Williams, M. A., Reddy, G., Quinn, M. J. Jr., and Johnson, M. S. Amsterdam: Elsevier, pp. 1–13.

Johnson, R. L., Anschutz, A. J., Smolen, J. M., Simcik, M. F., and R. L. Penn. 2007. The adsorption of perfluorooctane sulfonate onto sand, clay, and iron oxide surfaces. *Journal of Chemical & Engineering Data* 52:1165–1170.

Kalachova, K., Hradkova, P., Lankova, D., Hajslova, J., and J. Pulkrabova. 2012. Occurrence of brominated flame retardants in household and car dust from the Czech Republic. *Science of the Total Environment* 441:182–193.

Kannan, K., Franson, J. C., Bowerman, W. W., et al. 2001a. Perflurooctane sulfonate in fish-eating water birds including bald eagles and albatrosses. *Environmental Science and Technology* 35(15):3065–3070.

Kannan, K., Koistinen, J., Beckmen, K., et al. 2001b. Accumulation of perfluorooctane sulfonate in marine mammals. *Environmental Science and Technology* 35:1593–1598. doi: 10.1021/es001873w

Kannan, K., Choi, J. W., Iseki, N., et al. 2002a. Concentrations of perfluorinated acids in livers of birds from Japan and Korea. *Chemosphere* 49(3):225–231. doi: 10.1016/s0045-6535(02)00304-1

Kannan, K., Corsolini, S., Falandysz, J., et al. 2002b. Perfluorooctanesulfonate and related fluorinated hydrocarbons in marine mammals, fishes, and birds from coasts of the Baltic and the Mediterranean Seas. *Environmental Science and Technology* 36(15):3210–3216. doi: 10.1021/es020519q

Kannan, K., Hansen, K. J., Wade, T. L.., and J. P. Giesy. 2002c. Perfluorooctane sulfonate in oysters, *Crassostrea virginica*, from the Gulf of Mexico and the Chesapeake Bay, USA. *Archives of Environmental Contamination and Toxicology* 42(3):313–318. doi: 10.1007/s00244-001-0003-8

Kannan, K., Newsted, J., Halbrook, R. S., Giesy, J. P. 2002d. Perfluorooctanesulfonate and related fluorinated hydrocarbons in mink and river otters from the United States. *Environmental Science and Technology* 36, 2566–2571. doi: 10.1021/es0205028

Kannan, K., Corsolini, S., Falandysz, J., et al. 2004. Perfluorooctanesulfonate and related fluorochemicals in human blood from several countries. *Environmental Science and Technology* 38:4489–4495. doi: 10.1021/es0493446

Kannan, K., Tao, L., Sinclair, E., et al. 2005. Perfluorinated compounds in aquatic organisms at various trophic levels in a Great Lakes food chain. *Archives of Environmental Contamination and Toxicology* 48:559–566. doi: 10.1007/s00244-004-0133-x

Kannan, K., Perrotta, E., and N. J. Thomas. 2006. Association between perfluorinated compounds and pathological conditions in southern sea otters. *Environmental Science and Technology* 40:4943–4948.

Karnjanapiboonwong, A., Deb, S. K., Seenivasan, S., Wang, D., and T. A. Anderson. 2018. Perfluoroalkylsulfonic and carboxylic acids in earthworms (*Eisenia fetida*): Accumulation and effects results from spiked soils at PFAS concentrations bracketing environmental relevance. *Chemosphere* 199:168–173. doi: 10.1016/j.chemosphere.2018.02.027

Kawashima, Y., Uy-Yu, N., and H. Kozuka. 1989. Sex-related difference in the inductions by perfluorooctanoic acid of peroxisomal β-oxidation, microsomal 1-acylglycerophosphocholine acyltransferase and cytosolic long-chain acyl-CoA hydrolase in rat liver. *Biochemical Journal* 261:595–600.

Kawashima, Y., Matsunaga, T., Uy-Yu, N., and H. Kozuka. 1991. Induction by perfluorooctanoic acid of microsomal 1-acylglycerophosphocholine acyltransferase in rat kidney: Sex-related difference. *Biochemical Pharmacology* 42(10):1921–1926.

Kawashima, Y., Suzukii, S., Kozuka, H., Sato, M., and Y. Suzuki. 1994. Effects of prolonged administration of perfluorooctanoic acid on hepatic activities of enzymes which detoxify peroxide and xenobiotic in the rat. *Toxicology* 93:85–97.

Kawashima, Y., Kobayashi, H., Miura, H., and H. Kozuka. 1995. Characterization of hepatic responses of rat to administration of perfluorooctanoic and perfluorodecanoic acids at low levels. *Toxicology* 99:169–178.

Keil, D. E., Mehlmann, T., Butterworth, L., and M. M. Peden-Adams. 2008. Gestational exposure to per-fluorooctane sulfonate suppresses immune function in B6C3F1 mice. *Toxicological Sciences* 103:77–85.

Keller, J. M., Kannan, K., Taniyasu, S., et al. 2005. Perfluorinated compounds in the plasma of loggerhead and Kemp's Ridley sea turtles from the southeastern Coast of the United States. *Environmental Science and Technology* 39:9101–9108.

Kelly, B. C., Ikonomou, M. G., Blair, J. D., et al. 2009. Perfluoroalkyl contaminants in an arctic marine food web: Trophic magnification and wildlife exposure. *Environmental Science and Technology* 43(11):4037–4043.

Kennedy, G. L., Jr. 1985. Dermal toxicity of ammonium perfluorooctanoate. *Toxicology and Applied Pharmacology* 81:348–355.

Kennedy, G. L. Jr., Hall, G. T., Brittelli, M. R., Barnes, J. R., and H. C. Chen. 1986. Inhalation toxicity of ammonium perfluorooctanoate. *Food and Chemical Toxicology* 24(12):1325–1329.

Kennedy, G. L., Jr. 1987. Increase in mouse liver weight following feeding of ammonium perfluorooctanoate and related fluorochemicals. *Toxicology Letters* 39:295–300.

Kennedy, G. L., Jr., Butenhoff, G. L., Olsen, G. W., et al. 2004. The toxicology of perfluorooctanoate. *Critical Reviews in Toxicology* 34:351–384.

Kerstner-Wood, C., Coward, L., and G. Gorman. 2003. Protein binding of perfluorobutane sulfonate, perfluorohexane sulfonate, perfluorooctane sulfonate and perfluorooctanoate to plasma (human, rat, and monkey), and various human-derived plasma protein fractions. Birmingham, AL: Southern Research Institute. USEPA public docket, administrative record AR-226-1354.

Kim, S. and K. Kannan K. 2007. Perfluorinated acids in air, rain, snow, surface runoff, and lakes: Relative importance of pathways to contamination of urban lakes. *Environ Sci Technol* 41:8328–8334.

Kim, S. K., Shoeib, M., Kim, K. S., and J. E. Park. 2012. Indoor and outdoor poly- and perfluoroalkyl substances (PFASs) in Korea determined by passive air sampler. *Environmental Pollution* 162:144–150.

Kim, M., Son, J., Park, M. S., et al. 2013. *In vivo* evaluation and comparison of developmental toxicity and teratogenicity of perfluoroalkyl compounds using Xenopus embryos. *Chemosphere* 93(6):1153–1160.

Kim, M., Park, M. S., Son, J., et al. 2015. Pefluoroheptanoic acid affects amphibian embryogenesis by inducing the phosphorylation of ERK and JNK. *International Journal of Molecular Medicine* 36:1693–1700.

Kissa, E. 2001. *Fluorinated surfactants and repellents*, 2nd ed. Revised and expanded. New York: Marcel Dekker, Inc., pp. 1–101; 198–269; 349–379; 451–487.

Kowalczyk, J., Ehlers, S., Oberhausen, A., et al. 2013. Absorption, distribution, and milk secretion of the perfluoroalkyl acids PFBS, PFHxS, PFOS, and PFOA by dairy cows fed naturally contaminated feed. *Journal of Agricultural and Food Chemistry* 61(12):2903–2912.

Kubwabo, C., Stewart, B., Zhu, J., and L. Marro. 2005. Occurrence of perfluorosulfonates and other perfluorochemicals in dust from selected homes in the city of Ottawa, Canada. *Journal of Environmental Monitoring* 7(11):1074–1078.

Kudo, N., and Y. Kawashima. 2003. Induction of triglyceride accumulation in the liver of rats by perfluorinated fatty acids with different carbon chain lengths: Comparison with induction of peroxisomal beta-oxidation. *Biological and Pharmaceutical Bulletin* 26:47–51.

Kwadijk, C. J. A. F., Korytar, P., and A. A. Koelmans. 2010. Distribution of perfluorinated compounds in aquatic systems in The Netherlands. *Environmental Science and Technology* 44:3746–3751.

Kwadijk, C. J. A. F., Velzeboer, I., and A. A. Koelmans. 2013. Sorption of perfluorooctane sulfonate to carbon nanotubes in aquatic sediments. *Chemosphere* 90:1631–1636.

Labadie, P., and M. Chevreuil. 2011. Partitioning behaviour of perfluorinated alkyl contaminants between water, sediment and fish in the Orange River (nearby Paris, France). *Environmental Pollution* 159:391–397.

Langley, A. E., and G. D. Pilcher. 1985. Thyroid, bradycardic and hypothermic effects of perfluoro-n-decanoic acid in rats. *Journal of Toxicology and Environmental Health* 15:485–491.

Larson, E. S., Conder, J. M., and J. A. Arblaster. 2018. Modeling avian exposures to perfluoroalkyl substances in aquatic habitats impacted by historical aqueous film forming foam releases. *Chemosphere* 201:335–341.

Lasier, P. J., Washington, J. W., Hassan, S. M., and T. M. Jenkins. 2011. Perfluorinated chemicals in surface waters and sediments from northwest Georgia, USA, and their bioaccumulation in *Lumbriculus variegatus*. *Environmental Toxicology and Chemistry* 30:2194–2201.

Lau, C., Thibodeaux, J. R., Hanson, R. G., et al. 2003. Exposure to perfluorooctane sulfonate during pregnancy in rat and mouse. II: Post-natal evaluation. *Toxicological Sciences* 74:382–392.

Lau, C., Butenhoff, J. L., and J. M. Rogers. 2004. The developmental toxicity of perfluoroalkyl acids and their derivatives. *Toxicology and Applied Pharmacology* 198:231–241.

Lau, C., Thibodeaux, J. R., Hanson, R. G., et al. 2006. Effects of perfluorooctanoic acid exposure during pregnancy in the mouse. *Toxicological Sciences* 90(2):510–518.

Lau, C., Anitole, K., Hodes, C., et al. 2007. Perfluoroalkyl acids: A review of monitoring and toxicological findings. *Toxicological Sciences* 99(2):366–394.

Lau, C. 2012. Perfluoroalkyl acids: Recent research highlights. *Reproductive Toxicology* 33:405–409.

Lawlor, T.E. 1995. Mutagenicity Test with T-6342 in the Salmonella-Escherichia coli/ Mammalian-Microsome Reverse Mutation Assay. Laboratory Number: 17073-0-409. Corning Hazleton Inc., Vienna, VA. 3M Company, St. Paul, MN. U.S. Environmental Protection Agency Administrative Record 226-0436.

Levine, A. D., Libelo, E. L., Bugna, G., et al. 1997. Biogeochemical assessment of natural attenuation of JP-4-contaminated ground water in the presence of fluorinated surfactants. *Science of the Total Environment* 208(3):179–195.

Li, C., Ji, R., Schaffer, A., et al. 2012. Sorption of a branched nonylphenol and perfluorooctanoic acid on Yangtse river sediments and their model components. *Journal of Environmental Monitoring* 14:2653–2658.

Li, L., Zhai, Z., Liu, J., and J. Hu. 2015. Estimating industrial and domestic environmental releases of perfluorooctanoic acid and its salts in China from 2004 to 2012. *Chemosphere* 129:100–109.

Li, Y., Fletcher, T., Mucs, D., et al. 2017. Half-lives of PFOS, PFHxS and PFOA after end of exposure to contaminated drinking water *Occupational and Environmental Medicine* 75:46–51. doi: 10.1136/oemed-2017-104651

Liao, C., Wang, T., Cui, L., et al. 2009a. Changes in synaptic transmission, calcium current, and neurite growth by perfluorinated compounds are dependent on the chain length and functional group. *Environmental Science and Technology* 43:2099–2104.

Liao, C., Wang, T., Cui, L., et al. 2009b. Supporting information: Changes in synaptic transmission, calcium current, and neurite growth by perfluorinated compounds are dependent on the chain length and functional group. *Environmental Scienceand Technology* 43:2099–2104.

Lieder, P. H., York, R. G., Hakes, D. C., Chang, S. C., and J. L. Butenhoff. 2009. A two-generation oral gavage reproduction study with potassium perfluorobutanesulfonate (K+PFBS) in Sprague Dawley rats. *Toxicology* 259:33–45.

Lindim, C., van Gils, J., and I. T. Cousins. 2016. Europe-wide estuarine export and surface water concentrations of PFOS and PFOA. *Water Research* 103:124–132.

Liou, J. S-C., Szostek, B., DeRito, C. M., and E. L. Madsen. 2010. Investigating the biodegradability of perfluorooctanoic acid. *Chemosphere* 80:176–183.

Liu, R.C.M., M.E. Hurtt, J.C. Cook, and L.B. Biegel. 1996. Effect of the peroxisome proliferator, ammonium perfluorooctanoate (C8), on hepatic aromatase activity in adult male Crl:CD BR (CD) rats. *Fundamentals of Applied Toxicology* 30:220–228.

Liu, C., Chang, V. W., Gin, K. Y., and V. T. Nguyen. 2014. Genotoxicity of perfluorinated chemicals (PFCs) to the green mussel (*Perna viridis*). *Science of the Total Environment* 487:117–122.

Liu, Z., Lu, Y., Wang, T., et al. 2016. Risk assessment and source identification of perfluoroalkyl acids in surface and ground water: Spatial distribution around a mega-fluorochemical industrial park, China. *Environment International* 91:69–77.

Loccisano, A. E., Longnecker, M. P., Campbell, J. L. Jr., Andersen, M. E., and H. J. Clewell III. 2013. Development of PBPK models for PFOA and PFOS for human pregnancy and lactation life stages. *Journal of Toxicology and Environmental Health* A. 76(1):25–57; doi: 10.1080/15287394.2012.722523

Loewen, M., Halldorson, T., Wang, F., and G. Tomy. 2005. Fluorotelomer carboxylic acids and PFOS in rainwater from an urban center in Canada. *Environmental Science and Technology* 39:2944–2951. doi: 10.1021/es048635b

Loi, E. I. H., Yeung, L. W. Y., Taniyasu, S., et al. 2011. Trophic magnification of poly- and perfluorinated compounds in a subtropical food web. *Environmental Science and Technology* 45(13):5506–5513.

Long, Y., Wang, Y., Ji, G., Yan, L., Hu, F., and A. Gu. 2018. Neurotoxicity of perfluorooctane sulfonate to hippocampal cells in adult mice. *PLoS One* 8(1): e54176. doi: 10.1371/journal.pone.0054176

Lopez-Doval, S., Salgado, R., Pereiro, N., Moyano, R., and A. Lafuente. 2014. Perfluorooctane sulfonate effects on the reproductive axis in adult male rats. *Environmental Research* 134 (Special Issue):158–168. doi: 10.1016/j.envres.2014.07.006

Lu, L., Xu, L., Kang, T., and S. Cheng. 2012. Investigation of DNA damage treated with perfluoro-octane sulfonate (PFOS) on ZrO2/DDAB active nano-order film. *Biosensors and Bioelectronics* 35(1):180–185.

Lu, Y., Luo, B., Li, J., and J. Dai. 2015. Perfluorooctanoic acid disrupts the blood-testes barrier and activates TNFα/p38 MAPK signaling pathway *in vivo* and *in vitro*. *Archives of Toxicology* 90(4):971–983.

Luebker, D. J., Hansen, K. J., Bass, N. M., Butenhoff, J. L., and A. M. Seacat. 2002. Interactions of fluoro-chemicals with rat liver fatty acid-binding protein. *Toxicology* 176(3):175–185.

Luebker DJ, Case MT, York RG, Moore JA, Hanson KJ, and JL Butenhoff. 2005a. Two-generation reproduction and cross-foster studies of perfluorooctanesulfonate (PFOS) in rats. *Toxicology* 215: 126–148.

Luebker, D. J., York, R. G., Hansen, K. J., Moore, J. A., and J. L. Butenhoff. 2005b. Neonatal mortality from *in utero* exposure to perfluorooctanesulfonate (PFOS) in Sprague–Dawley rats: Dose-response, and biochemical and pharamacokinetic parameters. *Toxicology* 215:149–169.

Lupton, S. J., Huwe, J. L., Smith, D. J., Dearfield, K. L., and J. J. Johnston. 2012. Absorption and excretion of 14C-perfluorooctanoic acid (PFOA) in Angus cattle (*Bos taurus*). *Journal of Agricultural and Food Chemistry* 60:1128–1134.

Luster, M. I., Portier, C., Pait, D. G., et al. 1992. Risk assessment in immunotoxicology. I. Sensitivity and predictability of immune tests. *Fundamental and Applied Toxicology* 18(2):200–210.

Lv, Z., Li, G., Li, Y., et al. 2013. Glucose and lipid homeostasis in adult rat is impaired by early-life exposure to perfluorooctane sulfonate. *Environmental Toxicology* 28(9):532–542. doi: 10.1002/tox.20747

Macon, M. B., Villanueva, L. R., Tatum-Gibbs, K., et al. 2011. Prenatal perfluorooctanoic acid exposure in CD-1 mice: Low-dose developmental effects and internal dosimetry. *Toxicological Sciences* 122(1):134–145.

Mak, Y. L., Taniyasu, S., Yeung, L. W., et al. 2009. Perfluorinated compounds in tap water from China and several other countries. *Environmental Science and Technology* 43(13):4824–4829. doi: 10.1021/es900637a

Markoe, D. M. 1983. Primary skin irritation test with T-3371 in albino rabbits. Riker Laboratories. Study No. 0883EB0079, July 13, 1983. U.S. Environmental Protection Agency Administrative Record, 226-0424.

Martin, J. W., Mabury, S. A., Solomon, K. R., and D. C. Muir. 2003a. Bioconcentration and tissue distribution of perfluorinated acids in rainbow trout (*Oncorhynchus mykiss*). *Environmental Toxicology and Chemistry* 22:196–204.

Martin J. W., Mabury, S. A., Solomon, K. R., and D. C. G. Muir. 2003b. Dietary accumulation of perfluorinated acids in juvenile rainbow trout (*Oncorhynchus mykiss*). *Environmental Toxicology and Chemistry* 22:189–195.

Martin, J. W., Mabury, S. A., Solomon, K. R., and D. C. G. Muir. 2003c. Bioconcentration and tissue distribution of perfluorinated acids in rainbow (*Oncorhynchus mykiss*). *Environmental Toxicology and Chemistry* 22:196–204.

Martin, J. W., Whittle, D. M., Muir, D. C. G., and S. A. Mabury. 2004a. Perfluoroalkyl contaminants in a food web from Lake Ontario. *Environmental Science and Technology*
38(20):5379–5385.

Martin, J. W., Smithwick, M. M., Braune, B. M., et al. 2004b. Identification of long-chain
perfluo-rinated acids in biota from the Canadian Arctic. *Environmental Science and
Technology* 38(2):373–380. doi: 10.1021/es034727+

Martin, M. T., Brennan, R. J., Hu, W., et al. 2007. Toxicogenomic study of triazole fungicides
and perfluoro-alkyl acids in rat livers predict toxicity and categorizes chemicals based
on mechanisms of toxicity. *Toxicological Sciences* 97:595–613.

Marziali, L., Rosignoli, F., Valsecchi, S., Polesello, S., and F. Stefani. 2019. Effects of per-
fluoralkyl substances on a multigenerational scale: A case study with *Chironomus
riparius* (Diptera, Chironomidae). *Environmental Toxicology and Chemistry* 38(5):
988–999.

Meesters, R.J.W. and H.F. Schroder. 2004. Perfluorooctane sulfonate - a quite mobile anionic
anthropogenic surfactant, ubiquitously found in the environment. *Water Science and
Technology* 50(5):235–242.

Milinovic, J., Lacorte, S., Rigol, A., and M. Vidal. 2016. Sorption of perfluoroalkyl substances
in sewage sludge. *Environmental Science and Pollution Research* 23:8339–8348.

Miller, A., Elliott, J. E., Elliott, K. H., Lee, S., and F. Cyr. 2015. Temporal trends of perfluoroalkyl substances (PFAS) in eggs of coastal and offshore birds: Increasing PFAS levels
associated with offshore bird species breeding on the Pacific coast of Canada and win-
tering near Asia. *Environmental Toxicology and Chemistry* 34(8):1799–808.

Miralles-Marco, A., and S. Harrad. 2015. Perfluorooctane sulfonate: A review of human
exposure, biomonitoring and the environmental forensics utility of its chirality and
isomer distribution. *Environment International* 77:148–159.

Molina, E. D., Balander, R., Fitzgerald, S. D., et al. 2006. Effects of air cell injection of perfluorooctane sulfonate before incubation on development of the white leghorn chicken
(*Gallus domesticus*) embryo. *Environmental Toxicology and Chemistry* 25(1):227–32.

Moody, C. A., Kwan, W. C., Martin, J. W., Muir, D. C., and S. A. Mabury. 2001. Determination
of perfluorinated surfactants in surface water samples by two independent analytical techniques: Liquid chromatography/tandem mass spectrometry and ^{19}F NMR.
Analytical Chemistry 73(10):2200–2206.

Moody, C. A., Martin, J. W., Kwan, W. C., Muir, D. C., and S. A. Mabury. 2002. Monitoring
perfluor-inated surfactants in biota and surface water samples following an accidental release of fire-fighting foam into Etobicoke Creek. *Environmental Science and
Technology* 36(4):545–551.

Moody, C. A., and J. A. Field. 2003. Perfluorinated surfactants and environmental implications of their use in fire-fighting foams. *Environmental Science & Technology*
34(18):3864–3870.

Moody, C. A., Hebert, G. N., Strauss, S. H., and J. A. Field. 2003. Occurrence and persistence of perfluorooctanesulfonate and other perfluorinated surfactants in groundwater at a fire-training area at Wurtsmith Air Force Base, Michigan, USA. *Journal of
Environmental Monitoring* 5(2):341–345.

Morikawa, A., Kamei, N., Harada, K., et al. 2006. The bioconcentration factor of perfluo-
rooctane sulfonate is significantly larger than that of perfluorooctanoate in wild turtles
(*Trachemys scripta elegans* and *Chinemys reevesii*): An Ai River ecological study in
Japan. *Ecotoxicology and Environmental Safety* 65:14–21.

Morreale de Escobar, G., Obregon, M. J., Ruiz de Oña, C., and F. Escobar del Rey. 1988.
Transfer of thyroxine from the mother to the rat fetus near term: Effects on brain
3,5,3'-triiodothyronine deficiency. *Endocrinology* 122:1521–1531.

Morreale de Escobar, G., Obregon, M. J., and F. Escobar del Rey. 2004. Role of thyroid hormone during early brain development. *European Journal of Endocrinology* 151(Suppl 3):U25–U37.

Murli, H. 1995. Mutagenicity test on T-6342 in an in vivo mouse micronucleus assay. Vienna, VA: Corning Hazleton Inc. Study No. 17073-0-455, December 14, 1995. U.S. Environmental Protection Agency Administrative Record 1995, 226-0435.

Murli, H. 1996a. Mutagenicity test on T-6342 measuring chromosomal aberrations in human whole blood lymphocytes with a confirmatory assay with multiple harvests. Vienna, VA: Corning-Hazelton, Inc. Study No. 17073-0–449CO, November 1, 1996. U.S. Environmental Protection Agency Administrative Record 226-0433.

Murli, H. 1996b. Mutagenicity test on T-6564 measuring chromosomal aberrations in Chinese hamster ovary (CHO) cells with a confirmatory assay with multiple harvests. Vienna, VA: Corning Hazleton Inc. Study No. 17750-0-437CO, September 16, 1996. U.S. Environmental Protection Agency Administrative Record 226-0431.

Murli, H. 1996c. Mutagenicity test on T-6342 measuring chromosomal aberrations in Chinese hamster ovary (CHO) cells with a confirmatory assay with multiple harvests. Vienna, VA: Corning-Hazelton, Inc. Study No. 17073-0-437CO, September 16, 1996. U.S. Environmental Protection Agency Administrative Record 226-0434.

Murli, H. 1996d. Mutagenicity test on T-6564 in an *in-vivo* mouse micronucleus assay. Vienna, VA: Corning Hazleton Inc.. Study number 17750-0-455, November 1, 1996. U.S. Environmental Protection Agency Administrative Record 226-0430.

Nakayama, S., Strynar, M. J., Helfant, L., et al. 2007. Perfluorinated compounds in the Cape fear drainage basin in North Carolina. *Environmental Science and Technology* 41:5271–5276.

Nakayama, S. F., Strynar, M. J., Reiner, J. L., Delinsky, A. D., and A. B. Lindstrom. 2010. Determination of perfluorinated compounds in the Upper Mississippi river basin. *Environmental Science and Technology* 44:4103–4109.

Nakayama, S. F., Yoshikane, M., Onoda, Y., et al. 2019. Worldwide trends in tracing poly- and perfluoroalkyl substances (PFAS) in the environment. *Trends Analytical Chemistry* 121:115410. doi: 10.1016/j.trac.2019.02.011

Newsted, J. L., Jones, P. D., Coady, K., and J. P. Giesy. 2005. Avian toxicity reference values for perflurooctane sulfonate. *Environmental Science & Technology* 39:9357–9367.

Newsted, J. L., Beach, S. A., Gallagher, S. P., and J. P. Giesy. 2006. Pharmacokintics and acute lethality of perfluorooctane (PFSO) to juvenile mallard and northern bobwhite. *Archives of Environmental Contamination and Toxicology* 50:411–420.

Newsted, J. L., Coady, K. K., Beach, S. A., et al. 2007. Effects of perfluorooctane sulfonate on mallard and northern bobwhite quail exposed chronically via the diet. *Environmental Toxicology and Pharmacology* 23:1–9.

Newsted, J. L., Beach, S. A., Gallagher, S. P., and J. P. Giesy. 2008. Acute and chronic effects of perfluorobutane sulfonate (PFBS) on the mallard and northern bobwhite quail. *Archives of Environmental Contamination and Toxicology* 54:535–545.

Ngo, H. T., Hetland, R. B., Sabaredzovic, A., Haug, L. S., and I. L. Steffensen. 2014. In utero exposure to perfluorooctanoate (PFOA) or perfluorooctane sulfonate (PFOS) did not increase body weight or intestinal tumorigenesis in multiple intestinal neoplasia (Min/+) mice. *Environmental Research* 132:251–263.

Noker, P. E., and G. S. Gorman. 2003. *A pharmacokinetic study of potassium perfluorooctane sulfonate in the cynomolgus monkey.* Birmingham, AL: Southern Research Institute. USEPA public docket, administrative record AR-226-1356.

Nordén, M., Berger, U., and M. Engwall. 2016. Developmental toxicity of PFOS and PFOA in great cormorant (*Phalacrocorax carbo sinensis*), herring gull (*Larus argentatus*) and chicken (*Gallus gallus domesticus*). *Environmental Science and Pollution Research* 23(11):10855–10856.

NOTOX. 2000. Evaluation of the ability of T-7524 to induce chromosome aberrations in cultured peripheral human lymphocytes. NOTOX Project Number 292062. Hertogenbosch, The Netherlands.

NTN (National Toxics Network). 2015. NNT monograph on the persistence and toxicity of perfluorinated compounds in Australia. National Toxics Network (www.ntn.org.au). Lloyd-Smith, M., and Senjen, R., eds., pp. 13.

NTP (National Toxicology Program). 2005. Perfluorobutane sulfonate. Genetic toxicology – Bacterial mutagenicity. Chemical effects in biological systems (CEBS) – Study ID: A32037. Research Triangle Park, NC: National Institute of Environmental Sciences, National Institutes of Health.

NTP (National Toxicology Program). 2016. NTP monograph on immunotoxicity associated with exposure to perflurooctanoic acid (PFOA) or perflurooctane sulfonate (PFOS). National Toxicology Program, U.S. Dept. of Health and Human Services.

NTP (National Toxicology Program). 2019. NTP technical report on the toxicity studies of perfluoroalkyl carboxylates (perfluorohexanoic acid, perfluorooctanoic acid, perfluorononanoic acid, and perfluorodecanoic acid) administered by gavage to Sprague Dawley (Hsd:Sprague Dawley SD) rats. NTP monograph 97, National Toxicology Program, U.S. Dept. of Health and Human Services.

O'Brien, J. M., Crump, D., Mundy, L. J., et al. 2009. Pipping success and liver mRNA expression in chicken embryos exposed in ovo to C8 and C11 perfluorinated carboxylic acids and C10 perfluorinated sulfonate. *Toxicology Letters* 190:134–139.

Ochoa-Herrera, V., and R. Sierra-Alvarez. 2008. Removal of perfluorinated surfactants by sorption onto granular activated carbon, zeolite and sludge. *Chemosphere* 72: 1588–1593.

Oda, Y., Nakayama, S., Harada, K. H., and A. Koizumi. 2007. Negative results of the umu genotox-icity test of fluorotelomer alcohols and perfluorinated alkyl acids. *Environmental Health and Preventive Medicine* 12(5):217–219.

OECD. 2002. Hazard assessment of perfluorooctane sulfonate (PFOS) and its salts. Organisation for Economic Co-operation and Development. ENV/JM/RD(2002)17/FINAL. https://www.oecd.org/env/ehs/risk-assessment/2382880.pdf. Accessed December 17, 2020.

OECD. 2006. Lists of PFOS, PFAS, PFCA, and related compounds and chemicals that may degrade to PFCA: ENV/JM/MONO (2006)15. 2007. [http://www.oecd.org/LongAbstract/0,3425, en_2649_34375_39160347_119666_1_1_1,00.html]. Paris: Organisation for Economic Co-operation and Development.

Ohmori, K., Kudo, N., Katayama, K., and Y. Kawashima. 2003. Comparison of the toxicokinetics between perfluorocarboxylic acids with different carbon chain length. *Toxicology* 184:135–140.

Ojo, A.F., Peng C., Ng, J.C. 2020. Combined effects and toxicological interactions of perfluoroalkyl and polyfluoroalkyl substances mixtures in human liver cells (HepG2). *Environ. Pollut.* 262 (PtB):114182 DOI:10.1016/j.envpol.2020.114182.

Olsen, G. W., Burris, J. M., Burlew, M. M., and J. H. Mandel. 2003. Epidemiologic assessment of worker serum perfluorooctanesulfonate (PFOS), perfluorooctanoate (PFOA) concentrations, and medical surveillance examinations. *Journal of Occupational and Environmental Medicine/American College of Occupational and Environmental Medicine* 45(3):260–270.

Olsen, G. W., Mair, C. D., Church, T. R., et al. 2008. Decline in perfluorooctanesulfonate and other polyfluoroalkyl chemicals in American Red Cross adult blood donors, 2000–2006. *Environmental Science and Technology* 42:4989–4995.

Olsen, G. W., Chang, S. C., Noker, P. E., et al. 2009. A comparison of the pharmacokinetics of perfluorobutanesulfonate (PFBS) in rats, monkeys, and humans. *Toxicology* 256:65–74.

Olsen, G.W., et al. 2009. A comparison of the pharmacokinetics of perfluorobutanesulfonate (PFBS) in rats, monkeys, and humans. *Toxicology* 256(1–2):65–74. https://doi.org/10.1016/j.tox.2008.11.008.

O'Malley, K. D., and K. L. Ebbins. 1981. Repeat application 28-day percutaneous absorption study with T-2618CoC in albino rabbits. St Paul, MN: Riker Laboratories. U.S. Environmental Protection Agency Administrative Record 226-0446.

Onishchenko, N., Fischer, C., Ibrahim, W. N. W., et al. 2011. Prenatal exposure to PFOS or PFOA alters motor function in mice in a sex-related manner. *Neurotoxicity Research* 2011, 19, 452–461.

Organisation for Economic Cooperation and Development (OECD). 2002. Hazard assessment of perfluorooctanesulfonate (PFOS) and its salts. Unclassified ENV/JM/RD (2002)17/Final. Document No. JT00135607. Paris: Organisation for Economic Co-operation and Development.

Palmer, S. J., and H. O. Krueger. 2001. PFOS: A frog embryo teratogenesis assay –Xenopus (FETAX) Project No 454A-116. USEPA Docket AR226-1030a057. Easton, MD: Wildlife International, Ltd.

Pan, G., Jia, C., Zhao, D., et al. 2009. Effect of cationic and anionic surfactants on the sorption and desorption of perfluorooctane sulfonate (PFOS) on natural sediments. *Environmental Pollution* 157:325–330.

Peden-Adams, M. M., EuDaly, J. G., Dabra, S., et al. 2007. Suppression of humoral immunity following exposure to the perflu orinated insecticide sulfluramid. *Journal of Toxicology and Environmental Health* 70:1130–1141.

Peden-Adams, M. M., Keller, J. M., Eudaly, J. G., et al. 2008. Suppression of humoral immunity in mice following exposure to perfluorooctane sulfonate. *Toxicological Sciences* 104(1):144–154.

Peden-Adams, M. M., Stuckey, J. E., Gaworecki, K. M., et al. 2009. Developmental toxicity in white leghorn chickens following in ovo exposure to perfluorooctane sulfonate (PFOS). *Reproductive Toxicology* 27(3–4):307–318.

Perkins, R. G., Butenhoff, J. L., Kennedy , Jr., G. L., and M. J. Palazzolo. 2004. 13-Week dietary toxicity study of ammonium perfluorooctanoate (PFOA) in male rats. *Drug and Chemical Toxicology* 24(4):361–378.

Permadi, H., Lundgren, B., Andersson, K., and J. W. DePierre. 1992. Effects of perfluoro fatty acids on xenobiotic-metabolizing enzymes, enzymes which detoxify reactive forms of oxygen and lipid peroxidation in mouse liver. *Biochemical Pharmacology* 44(6):1183–1191.

Perra, G. M., Focardi, S. E., and C. Guerranti. 2013. Levels and spatial distribution of perfluorinated compounds (PFCs) in superficial sediments from the marine reserves of the Tuscan Archipelago National Park (Italy). *Marine Pollution Bulletin* 13(76):379–382.

Peterson, R. E. 1990. Toxicology of perfluorodecanoic acid. U.S. Air Force Office of Scientific Research. Final Technical Report No. AFOSR-85-0207.

Pinkas, A., Slotkin, T. A., Brick-Turin, Y., Van der Zee, E. A., and J. Yanai. 2010. Neurobehavioral teratogenicity of perfluorinated alkyls in an avian model. *Neurotoxicology and Teratology* 32:182–186.

Post, G. B., Cohn, P. D., and K. R. Cooper. 2012. Perfluorooctanoic acid (PFOA), an emerging drinking water contaminant: A critical review of recent literature. *Environmental Research* 116:93–117. doi: 10.1016/j.envres.2012.03.007

Post, G. B., Louis, J. B., Lippincott, R. L., and N. A. Procopio. 2013. Occurrence of perfluorinated compounds in raw water from New Jersey public drinking water systems. *Environmental Science and Technology* 47:13266–13275.

Prevedouros, K., Cousins, I. T., Buck, R. C., and S. H. Korzeniowski. 2006. Sources, fate and transport of perfluoro-carboxylates. *Environmental Science and Technology* 40(1):32–44. doi: 10.1021/es0512475

Qazi, M. R., Abedi, M. R., Nelson, B. D., DePierre, J. W., and M. Abedi-Valugerdi. 2010. Dietary exposure to perfluorooctanoate or perfluorooctane sulfonate induces hypertrophy in centrilobular hepatocytes and alters the hepatic immune status in mice. *International Immunopharmacology* 10:1420–1427.

Qazi, M. R., Nelson, B. D., Depierre, J. W., and M. Abedi-Valugerdi. 2010. 28-Day dietary exposure of mice to a low total dose (7 mg/kg) of perfluorooctanesulfonate (PFOS) alters neither the cellular compositions of the thymus and spleen nor humoral immune responses: Does the route of administration play a pivotal role in PFOS-induced immunotoxicity? *Toxicology* 267:132–139.

Qazi, M. R., Nelson, B. D., DePierre, J. W., and M. Abedi-Valugerdi. 2012. High-dose dietary exposure of mice to perfluorooctanoate or perfluorooctane sulfonate exerts toxic effects on myeloid and B-lymphoid cells in the bone marrow and these effects are partially dependent on reduced food consumption. *Food and Chemical Toxicology* 50:2955–2963.

Qu, Y., Zhang, C., Li, F., et al. 2009. Equilibrium and kinetics study on the adsorption of perfluorooctanoic acid from aqueous solution onto powdered activated carbon. *Journal of Hazardous Materials* 169:146–152.

Quinete, N., Orata, F., Maes, A., et al. 2010. Degradation studies of new substitutes for perfluorinated surfactants. *Archives of Environmental Contamination and Toxicology* 59(1):20–30.

Quist, E. M., Filgo, A. J., Cummings, C. A., Kissling, G. E., and M. J. Hoenerhoff. 2015. Hepatic mitochondrial alteration in CD-1 mice associated with prenatal exposures to low doses of perfluorooctanoic acid (PFOA). *Toxicologic Pathology* 41:546–557.

Rahman, M. F., Peldszus, S., and W. B. Anderson. 2014. Behaviour and fate of perfluoroalkyl and poly fluoroalkyl substances (PFASs) in drinking water treatment: A review. *Water Research* 50:318–340.

Renner, R. 2006. The long and the short of perfluorinated replacements. *Environmental Science and Technology* 40:12–13.

Rigden, M., Pelletier, G., Poon, R., et al. 2015. Assessment of Urinary Metabolite Excretion After Rat Acute Exposure to Perfluorooctanoic Acid and Other Peroxisomal Proliferators. *Archives of Environmental Contamination and Toxicology* 68(1): 148–158.

Roland, K., Kestemont, P., Loos, R., et al. 2014. Looking for protein expression signatures in European eel peripheral blood mononuclear cells after *in-vivo* exposure to perfluorooctane sulfonate and a real world field study. *Science of the Total Environment* 468–469:958–67.

Rosal, R., Rodea-Palomares, I., Boltes, K., et al. 2010. Ecotoxicological assessment of surfactants in the aquatic environment: Combined toxicity of docusate sodium with chlorinated pollutants. *Chemosphere* 81:288–293.

Route, W. T., Russell, R. E., Lindstrom, A. B., Strynar, M. J., and R. L. Key. 2014. Spatial and temporal patterns in concentrations of perfluorinated compounds in bald eagle nestlings in the upper Midwestern United States. *Environmental Science & Technology* 48(12):6653–60.

Rusch, G. M., Rinehart, W. E., and C. A. Bozak. 1979. An acute inhalation toxicity study of T-2306 CoC in the rat. Project 78–7185. Bio/dynamics, Inc.

Sadhu, D. 2002. CHO/HGPRT forward mutation assay – ISO (T6.889.7). Bedford, MA: Toxicon Corporation. Report No. 01-7019-G1, March 28, 2002. U.S. Environmental Protection Agency Administrative Record 226-1101.

Saito, N., Sasaki, K., Nakatome, K., et al. 2003. Perfluorooctane sulfonate concentrations in surface water in Japan. *Archives of Environmental Contamination and Toxicology* 45:149–158.

Saito, N., Harada, K., Inoue, K., et al. 2004. Perfluorooctanoate and perfluorooctane sulfonate concentrations in surface waters in Japan. *Journal of Occupational Health* 46:49–59. doi: 10.1539/joh.46.49

Sato, I., Kawamoto, K., Nishikawa, Y., et al. 2009. Neurotoxicity of perflurooctane sulfonate (PFOS) in rats and mice after single oral exposure. *Journal of Toxicological Sciences* 34:569–574.

Saunders, B. C. 1972. *Carbon-fluorine compounds; chemistry, biochemistry, and biological activities.* Eds. Elliott, K., and Birch, J. Amsterdam: Associated Scientific Publishers.

Savu, P. 2000. *Fluorine-containing polymers, perfluoroalkanesulfonic acids. Kirk-Othmer encyclopedia of chemical technology (1999-2016).* New York: John Wiley & Sons.

Schlummer, M., Gruber, L., Fiedler, D., Kizlauskas, M., and J. Müller. 2013. Detection of fluoro-telomer alcohols in indoor environments and their relevance for human exposure. *Environment International*, 57–58:42–49.

Seacat, A. M., Thomford, P. J., Hansen, K. J., et al. 2002. Sub-chronic toxicity studies on perfluorooctanesulfonate potassium salt in cynomolgus monkeys. *Toxicological Sciences* 68:249–264.

Sepulvado, J. G., Blaine, A. C., Hundal, L. S., and C. P. Higgins. 2001. Occurrence and fate of perfluorochemicals in soil following the land application of municipal biosolids. *Environmental Science and Technology* 45(19):8106–8112.

Sheng, N., Zhou, X., Zheng, F., et al. 2017. Comparative hepatotoxicity of 6:2 fluorotelomer carboxylic acid and 6:2 fluorotelomer sulfonic acid, two fluorinated alternatives to long-chain perfluoroalkyl acids, on adult male mice. *Archives of Toxicology* 91:2909–2919.

Shi, Z., Zhang, H., Liu, Y., Xu, M., and J. Dai. 2007. Alterations in gene expression and testosterone synthesis in the testes of male rats exposed to perfluorododecanoic acid. *Toxicological Sciences* 98:206–215.

Shipley, J. M., Hurst, C. H., Tanaka, S. S., et al. 2004. Trans-activation of PPARalpha and induction of PPARalpha target genes by perfluorooctane-based chemicals. *Toxicological Sciences* 80(1):151–160.

Shoeib, M., Harner, T., and P. Vlahos. 2006. Perfluorinated chemicals in the arctic atmosphere. *Environmental Science and Technology* 40:7577–7583.

Shoeib, M., Harner, T., Webster, G. M., and S. C. Lee. 2011. Indoor sources of poly- and perfluorinated compounds (PFCS) in Vancouver, Canada: Implications for human exposure. *Environmental Science and Technology* 45(19):7999–8005.

Simcik, M. F. 2005. Global transport and fate of perfluorochemicals. *Journal of Environmental Monitoring* 7:759–763.

Sinclair, E., Mayack, D. T., Roblee, K., Yamashita, N., and K. Kannan. 2006. Occurrence of perfluoroalkyl surfactants in water, fish, and birds from New York State. *Archives of Environmental Contamination and Toxicology* 50:398–410. doi: 10.1007/s00244-005-1188-z

Skakkebaek, N. E., Rajpert-De Meyts, E., and K. M. Main. 2001. Testicular dysgenesis syndrome: An increasingly common developmental disorder with environmental aspects. *Human Reproduction* 16:972–978.

Smithwick, M., Mabury, S. A., Solomon, K. R., et al. 2005a. Circumpolar study of perfluoroalkyl contaminants in polar bears (*Ursus maritimus*). *Environmental Science and Technology* 39(15):5517–5523. doi: 10.1021/es048309w

Smithwick, M., Muir, D. C., Mabury, S. A., et al. 2005b. Perfluoroalkyl contaminants in liver tissue from East Greenland polar bears (*Ursus maritimus*). *Environmental Toxicology and Chemistry* 24(4):981–986. doi: 10.1897/04-258r.1

Smithwick, M., Norstrom, R. J., Mabury, S. A., et al. 2006. Temporal trends of perfluoro-alkyl contaminants in polar bears (*Ursus maritimus*) from two locations in the North American arctic, 1972-2002. *Environmental Science and Technology* 40:1139–1143. doi: 10.1021/es051750h

Smits, J. E. and S. Nain. 2013. Immunomodulation and hormonal disruption without com-promised disease resistance in perfluorooctanoic acid (PFOA) exposed Japanese quail. *Environmental Pollution* 179:13–8.

So, M. K., Taniyasu, S., Lam, P. K. S., et al. 2006. Alkaline digestion and solid phase extrac-tion method for perfluorinated compounds in mussels and oysters from South China and Japan. *Archives of Environmental Contamination and Toxicology* 50:240–248. doi: 10.1007/s00244-005-7058-x

Sohlenius, A.-K., Lundgren, B., and J. W. DePierre. 1992. Perfluorooctanoic acid has persis-tent effects on peroxisome proliferation and related parameters in mouse liver. *Journal of Biochemical Toxicology* 7(4):205–212.

Sohlenius, A. K., Eriksson, A. M., Högström, C., Kimland, M., and J. W. DePierre. 1993. Perfluorooctane sulfonic acid is a potent inducer of peroxisomal fatty acid beta-oxi-dation and other activities known to be affected by peroxisome proliferators in mouse liver. *Pharmacology and Toxicology* 72(2):90–93.

Son, H.-Y., Kim, S. H., Shin, H. –I., Bae, H. I., and J.- H. Yang. 2008. Perfluorooctanoic acid-induced hepatic toxicity following 21-day oral exposure in mice. *Archives of Toxicology* 82:239–246.

Son, H.-Y., Lee, S., Tak, E. N., et al. 2009. Perfluorooctanoic acid alters T lymphocyte phenotypes and cytokine expression in mice. *Environmental Toxicology* 24(6):580–588.

Stahl, T., Falk, S., Failing, K., et al. 2012. Perfluorooctanoic acid and perfluorooctane sul-fonate in liver and muscle tissue from wild boar in Hesse, Germany. *Archives of Environmental Contaminationa and Toxicology* 62(4):696–703.

Stasinakis, A., Petalas, A. V., Mamia, D., and N. S. Thomaidis. 2008. Application of the OECD 301F respirometric test for the biodegradability assessment of various poten-tial endocrine disrupting chemicals. *Bioresource Technology* 99(9):3458–3467. doi: 10.1016/j.biortech.2007.08.002

Sturm, R., and L. Ahrens. 2010. Trends of polyfluoroalkyl compounds in marine biota and in humans. *Environmental Chemistry* 7(6):457–484.

Staples, R. E., Burgess, B. A., and W. D. Kerns. 1984. The embryo-fetal toxicity and terato-genic potential of ammonium perfluorooctanoate (PFOA) in the rat. *Fundamental and Applied Toxicology* 4(3):429–440.

Starkov, A. A., and K. B. Wallace. 2002. Structural determinants of fluorochemical-induced mitochondrial dysfunction. *Toxicological Sciences* 66(2):244–252.

Stock, N. L.., Lau, F. K., Ellis, D. A., et al. 2004. Polyfluorinated telomer alcohols and sul-fonamides in the North American troposphere. *Environmental Science and Technology* 38:991–996.

Stock, N. L., Furdui, V. I., Muir, D. C. G., and S. A. Mabury. 2007. Perfluoroalkyl contami-nants in the Canadian Arctic: Evidence of atmospheric transport and local contamina-tion. *Environmental Science and Technology* 41:3529–3536.

Strömqvist, M., Olsson, J. A., Kärrman, A., and B. Brunström. 2012. Transcription of genes involved in fat metabolism in chicken embryos exposed to the peroxisome prolifera-tor-activated receptor alpha (PPARα) agonist GW7647 or to perfluorooctane sulfonate (PFOS) or perfluorooctanoic acid (PFOA). *Comparative Biochemistry and Physiology, Part C* 156:29–36.

Strum, R., and L. Ahrens. 2011. Trends of polyfluoroalkyl compounds in marine biota and in humans. *Environmental Chemistry* 7:457–484.

Strynar, M. J., Lindstrom, A. B., Nakayama, S. F., Egeghy, P. P., and L. J. Helfant. 2012. Pilot scale application of a method for the analysis of perfluorinated compounds in surface soils. *Chemosphere* 86(3):252–257.

Suh, C. H., Cho, N. K., Lee, C. K., et al. 2011. Perfluorooctanoic acid-induced inhibition of placental prolactin-family hormone and fetal growth retardation in mice. *Molecular and Cellular Endocrinology* 337:7–15.

Sundstrom, M., Chang, S. C., Noker, P. E., et al. 2012. Comparative pharmacokinetics of perfluorohexanesulfonate (PFHxS) in rats, mice, and monkeys. *Reproductive Toxicology* 33:441–451.

Tan, X., Xie, G., Sun, X., et al. 2013. High-fat diet feeding exaggerates perfluorooctanoic acid-induced liver injury in mice via modulating multiple metabolic pathways. *PLOS One* 8(4):e61409.

Taniyasu, S., Yamashita, N., Moon, H. B., et al. 2013. Does wet precipitation represent local and regional atmospheric transportation by perfluorinated alkyl substances? *Environment International* 55C:25–32.

Tao, L., Kannan, K., Kajiwara, N., et al. 2006. Perfluorooctanesulfonate and related fluorochemicals in albatrosses, elephant seals, penguins, and polar skuas from the Southern Ocean. *Environmental Science and Technology* 40:7642–7648.

Thibodeaux, J. R., Hanson, R. G., Rogers, J. M., et al. 2003. Exposure to perfluorooctane sulfonate during pregnancy in rat and mouse. I: Maternal and prenatal evaluations. *Toxicological Sciences* 74:369–381.

Thomford, P. J. 2002. 104–week dietary chronic toxicity and carcinogenicity study with perflorooctane sulfonic acid potassium salt (PFOS); T-6295) in rats. Final Report, 3MT-6295 (Covance Study No 6329-183) Volumes I-IX. January 2, 2002. 3M St. Paul MN, pp. 4068.

Thompson, W. 2018. Chronic toxicity of perfluoroheptanoic acid (PFHpA) and perfluorooctanoic acid (PFOA) to northern bobwhite (*Colinus virginianus*). Texas Tech University, Doctoral Dissertation 2018. Available at: https://ttu-ir.tdl.org/bitstream/handle/2346/74514/THOMPSON-THESIS-2018.url?sequence=4&isAllowed=y.

Thottassery, J., Winberg, L., Youssef, J., et al. 1992. Regulation of perfluorooctanoic acid-induced peroxisomal enzyme activities and hepatocellular growth by adrenal hormones. *Hepatology* 15(2): 316–322.

Tittlemier, S. A., Pepper, K., Seymour, C., et al. 2007. Dietary exposure of Canadians to perfluorinated carboxylates and perfluorooctane sulfonate via consumption of meat, fish, fast foods, and food items prepared in their packaging. *Journal of Agricultural and Food Chemistry* 55(8):3203–3210. doi: 10.1021/jf0634045

Trier, X., Granby, K., Christensen, J. H. 2011. Polyfluorinated surfactants (PFS) in paper and board coatings for food packaging. *Environmental Science and Pollution Research* 18(7):1108–1120.

Tucker, D. E., Macon, M. B., Strynar, et al. 2015. The mammary gland is a pensitive pubertal target in CD-1 and C57BL/6 mice following perinatal perfluorooctamoic acid (PFOA) exposure. *Reproductive Toxicology* 54:26–36.

Tue, N. M., Takahashi, S., Suzuki, G., et al. 2013. Contamination of indoor dust and air by polychlorinated biphenyls and brominated flame retardants and relevance of non-dietary exposure in Vietnamese informal e-waste recycling sites. *Environment International* 51:160–167.

U.K. Environment Agency. 2004. Environmental risk evaluation report: Perfluorooctanesulphonate (PFOS). ISBN: 978-1-84911-124-9. Available at: https://www.gov.uk/government/uploads/system/uploads/attachment_data/file/290857/scho1009brbl-e-e.pdf.

United Nations Environment Programme (UNEP). 2006. Risk profile on perfluorooctane sulfonate. Geneva: Stockholm Convention on Persistent Organic Pollutants Review Committee. 6–10 November 2006.

United Nations Environment Programme (UNEP). 2007. Risk management evaluation on perfluorooctane sulfonate. Stockholm Convention on Persistent Organic Pollutants Review Committee. Geneva: United Nations Enviornemntal Pogramme. 19–23, 2007.

U.S. Army Center for Health Promotion and Preventive Medicine (USACHPPM). 2000. Standard practice for wildlife toxicity reference values. Technical Guide 254. Prepared by Johnson, M. S. and McAtee, M. J. Aberdeen Proving Ground, MD: U.S. Army Center for Health Promotion and Preventive Medicine.

U.S. Environmental Protection Agency (USEPA). 1998. Health effects test guidelines: OPPTS 870.7800 immunotoxicity. USEPA/712/C-98/351. Washington, DC: Office of Prevention Pesticides and Toxic Substances. 1–11. Available at: http://www.USEPA.gov /ocspp/pubs/frs/publications/Test_Guidelines/series870.htm.

U.S. Environmental Protection Agency (USEPA). 2002. Revised draft hazard assessment of perfluorooctanoic acid and its salts. Washington, DC: U.S. Environmental Protection Agency, Office of Pollution Prevention and Toxics, Risk Assessment Division, pp. 103.

U.S. Environmental Protection Agency (USEPA). 2003. Preliminary risk assessment of the developmental toxicity associated with exposure to perfluorooctanoic acid and its salts. Washington, DC U.S. Environmental Protection Agency Office of Pollution Prevention and Toxics Risk Assessment Division.

U.S. Environmental Protection Agency (USEPA). 2006. Assessing and managing chemicals under TSCA. 2010/2015 PFOA stewardship program. Available at: http://www.USEPA .gov/opptintr/pfoa/pubs/stewardship [Last updated: 19 May 2016].

U.S. Environmental Protection Agency (USEPA). 2009. Long-chain perfluorinated chemicals (PFCs) action plan. 2009. Available at: http://www.USEPA.gov/opptintr/existingch emicals/pubs/actionplans/pfcs.html.

U.S. Environmental Protection Agency (USEPA). 2014. National Center for Environmental Assessment, Office of Research and Development. Provisional peer-reviewed toxicity values for perfluorobutane sulfonate (CASRN 375-73-5) and related compound potassium perfluorobutane sulfonate (CASRN 29420-49-3). Final report 07-17-2014.

U.S. Environmental Protection Agency (USEPA). 2014a. Draft health effects document for perfluorooctane acid (PFOA). Office of Water. Available at: https://peerreview.versar .com/USEPA/pfoa/pdf/Health-Effects-Document-for-Perfluorooctanoic-Acid-(PFOA). pdf.

U.S. Environmental Protection Agency (USEPA). 2014b. Emerging contaminants – Perfluoro-octane sulfonate (PFOS) and perfluorooctanoic acid (PFOA). Available at: http://www 2.USEPA.gov/sites/production/files/2014-04/documents/factsheet_contaminant_pfos_ pfoa_march2014.pdf.

U.S. Environmental Protection Agency (USEPA). 2016a. Health effects support document for perfluorooctane sulfonate (PFOS). USEPA 822-R-16-002. Washington, DC: Office of Water.

U.S. Environmental Protection Agency (USEPA). 2016b. Health effects support document for perfluorooctane acid (PFOA). USEPA 822-R-16-003. Washington, DC: Office of Water.

Uy-Yu, N., Kawasjoma, Y., and H. Kozuka. 1990a. Comparative studies on sex-related difference in biochemical responses of livers to perfluorooctanoic acid between rats and mice. *Biochemical Pharmacology* 39(9):1492–1495.

Uy-Yu, N., Kawasjoma, Y., and H. Kozuka. 1990b. Effects of chronic administration of perfluorooctanoic acid on fatty acid metabolism in rat liver: Relationship among stearoyl-coenzyme A desaturase, 1-acylglycerophosphocholine acyltransferase, and acyl composition of microsomal phospatidylcholine. *Journal of Pharmacobio-Dynamics* 13:581–590.

Van Rafelghem, M. J., Inhorn, S. L., and R. E. Peterson. 1987. Effects of perfluorodecanoic acid on thyroid status in rats. *Toxicology and Applied Pharmacology* 87:430–439.

Verreault, J., Berger, U., and G. W. Gabrielsen. 2007. Trends of perfluorinated alkyl substances in herring gull eggs from two coastal colonies in Northern Norway: 1983–2003. *Environmental Science and Technology* 41(19):6671–6677.

Vetvicka, V., and J. Vetvickova. 2013. Reversal of perfluorooctane sulfonate-induced immunotoxicity by a glucan–resveratrol–vitamin C combination. *Oriental Pharmacy and Experimental Medicine* 13(1):77–84.

Viberg H, Lee I, Eriksson P. 2013. Adult dose-dependent behavioral and cognitive disturbances after a single neonatal PFHxS dose. *Toxicology* 304;185–191. https://doi.org/10.1016/j.tox.2012.12.013.

Vicente, J., Sanpera, C., García-Tarrasón, M., Pérez, A., and S. Lacorte. 2015. Perfluoroalkyl and polyfluoroalkyl substances in entire clutches of Audouin's gulls from the Ebro Delta. *Chemosphere* 119 Suppl:S62–568.

Vierke, L., Staude, C., Biegel-Englewr, A., Drost, W., and C. Schulte. 2012. Perfluorooctanoic acid (PFOA) — main concerns and regulatory developments in Europe from an environmental point of view. *Environmental Sciences Europe* 24:16. doi: 10.1186/2190-4715-24-16

Wang, N., J. Liu, R.C. Buck, S.H. Korzeniowski, B.W. Wolstenholme, P.W. Folsom, and L.M. Sulecki. 2011. 6:2 fluorotelomer sulfonate aerobic biotransformation in activated sludge of waste water treatment plants. *Chemosphere* 82 (6):853–8.

Wan, H. T., Zhao, Y. G., Wei, X., et al. 2012. PFOS-induced hUSEPAtic steatosis, the mechanistic actions on β-oxidation and lipid transport. *Biochimica et Biophysica Acta* 1820(7):1092–101.

Wang, F., Liu, W., and Y. Jin. 2011. Interaction of PFOS and BDE-47 co-exposure on thyroid hormone levels and TH-related gene and protein expression in developing rat brains. *Toxicological Sciences* 121:279–291.

Wang, F., Liu, W., Jin, Y., Wang, F., and J. Ma. 2015a. Prenatal and neonatal exposure to perfluorooctane sulfonic acid results in aberrant changes in miRNA expression profile and levels in developing rat livers. *Environmental Toxicology* 30:712–723. doi: 10.1002/tox.21949

Wang, F., Shih, K., Ma, R., and X-. Y. Li. 2015b. Influence of cations on the partition behavior of perfluoroheptanoate (PFHpA) and perfluorohexanesulfonate (PFHxS) on wastewater sludge. *Chemosphere* 131:178–183.

Wang, L., Wang, Y., Liang, Y., et al. 2014. PFOS induced lipid metabolism disturbances in BALB/c mice through inhibition of low density lipoproteins excretion. *Scientific Reports* 3(4):4582. doi: 10.1038/srep04582

Wang, Y., Yeung, L. W. Y., Taniyasu, S., et al. 2008. Perfluorooctane sulfonate and other fluoro-chemicals in waterbird eggs from south China. *Environmental Science and Technology* 42(21):8146–8151. doi: 10.1021/es8006386

Wang, Y., Zhang, X., Wang, M., et al. 2015. Mutagenic effects of perfluorooctanesulfonic acid in gpt delta transgenic system are mediated by hydrogen peroxide. *Environmental Science and Technology* 49(10):6294–6303.

Wang, Z., Cousins, I. T., Scheringer, M., Buck, R. C., and K. Hungerbuhler. 2014. Global emission inventories for C4-C14 perfluoroalkyl carboxylic acid (PFCA) homologues from 1951 to 2030, Part I: Production and emissions from quantifiable sources. *Environment International* 70:62–75.

Wania, F. 2007. A global mass balance analysis of the source of perfluorocarboxylic acids in the Arctic Ocean. *Environmental Science and Technology* 41:4529–4535.

Wei, S., Chen, L. Q., Taniyasu, S., et al. 2007. Distribution of perfluorinated compounds in surface seawaters between Asia and Antarctica. *Marine Pollution Bulletin* 54:1813–1838. doi: 10.1016/j.marpolbul.2007.08.002

Wieneke, N., Neuschäfer-Rube, F., Bode, L. M., et al. 2009. Synergistic acceleration of thyroid hormone degradation by phenobarbital and the peroxisome proliferation receptor agonist WY14643. *Toxicology and Applied Pharmacology* 240:99–107. doi: 10.1016/j.taap.2009.07.014

White, S. S., Calafat, A. M., Kuklenyik, Z., et al. 2007. Gestational PFOA exposure of mice is associated with altered mammary gland development in dams and female offspring. *Toxicological Sciences* 96(1):133–144.

White, S. S., Kato, K., Jia, L. T., et al. 2009. Effects of perfluorooctanoic acid on mouse mammary gland development and differentiation resulting from cross-foster and restricted gestational exposures. *Reproductive Toxicology* 27:289–298.

White, S. S., Fenton, S. E., and E. P. Hines. 2011a. Endocrine disrupting properties of perfluoro-octanoic acid. *Journal of Steroid Biochemistry and Molecular Biology* 127:16–26.

White, S. S., Stanko, J. P., Kato, K., et al. 2011b. Gestational and chronic low-dose PFOA exposures and mammary gland growth and differentiation in three generations of CD-1 mice. *Environmental Health Perspectives* 119(8):1070–1076.

Wildlife International Ltd. 2001. PFBS: An activated sludge respiration test. Wildlife International Ltd., Project No. 454E-102A.

Wilkins, P. 2001. Perflurooctanesulfonate, potassium salt (PFOS): Acute contact toxicity study with honeybee. Study number HT5601. USEPA docket AR226-1018 – Environmental biology Group. Sand Hutton, York, UK: Central Science Laboratory.

Wolf, C. J., Fenton, S. E., Schmid, J. E., et al. 2007. Developmental toxicity of perfluorooctanoic acid in the CD-1 mouse after cross-foster and restricted gestational exposure. *Toxicological Sciences* 95:462–473.

Wolf, D. C., Moore, T., Abbott, B. D., et al. 2008. Comparative hepatic effects of perfluorooctanoic acid and WY 14,643 in PPAR-a knockout and wild-type mice. *Toxicological Pathology* 36:632–639.

Wolf, C. J., Zehr, R. D., Schmid, J. E., Lau, C., and B. D. Abbott. 2010. Developmental effects of perfluorononanoic acid in the mouse are dependent on peroxisome proliferator-activated receptor-alpha. *PPAR Research*. 2010:282896, 11 pp. doi: 10.1155/2010/282896.

World Health Organization (WHO). 2012. Guidance for immunotoxicity risk assessment for chemicals. International Programme on Chemical Safety (IPCS), Harmonization Project No. 10. Geneva. 2012. Available at: http://www.inchem.org/documents/harmproj/harmproj/harmproj10.pdf.

Xie, S., Wang, T., Liu, S., et al. 2013. Industrial source identification and emission estimation of perfluorooctane sulfonate in China. *Environment International* 52:1–8.

Xie, Y., Yang, Q., Nelson, B. D., and J. W. DePierre. 2002. Characterization of the adipose tissue atrophy induced by peroxisome proliferators in mice. *Lipids* 37(2):139–146.

Xie, Y., Yang, Q., Nelson, B. D., and J. W. DePierre. 2003. The relationship between liver peroxisome proliferation and adipose tissue atrophy induced by peroxisome proliferator exposure and withdrawal in mice. *Biochemical Pharmacology* 66:749–756.

Xing, J., Wang, G., Zhao, J., et al. 2006. Toxicity assessment of perfluorooctane sulfonate using acute and sub-chronic male C57BL/6J mouse models. *Environmental Pollution* 210:388–396.

Xu, D., Li, C., Wen, Y., and W. Liu. 2013. Antioxidant defense system responses and DNA damage of earthworms exposed to perfluorooctane sulfonate (PFOS). *Environmental Pollution* 174:121–127.

Yahia, D., Tsukuba, C., Yoshida, M., Sato, I., and S. Tsuda. 2008. Neonatal death of mice treated with perfluorooctane sulfonate. *The Journal of Toxicological Sciences* 33:219–226.

Yahia, D., El-Nasser, M. A., Abedel-Latif, M., et al. 2010. Effects of perfluorooctanoic acid (PFOA) exposure to pregnant mice on reproduction. *Journal of Toxicological Sciences* 35(4):527–533.

Yamashita, N., Kannan, K., Taniyasu, S., et al. 2004. Analysis of perfluorinated acids at parts-perquadrillion levels in seawater using liquid chromatography-tandem mass spectrometry. *Environmental Science and Technology* 38:5522–5528. doi: 10.1021/es0492541

Yamashita, N., Kannan, K., Taniyasu, S., et al. 2005. A global survey of perfluorinated acids in oceans. *Marine Pollution Bulletin* 51:658–668. doi: 10.1016/j.marpolbul.2005.04.026

Yamashita, N., Taniyasu, S., Petrick, G., et al. 2008. Perfluorinated acids as novel chemical tracers of global circulation of ocean waters. *Chemosphere* 70:1247–1255. doi: 10.1016/j.chemosphere.2007.07.079

Yanai, J., Dotan, S., Goz, R., et al. 2008. Exposure of developing chicks to perfluorooctanoic acid induces defects in prehatch and early posthatch development. *Journal of Toxicology and Environmental Health, Part A* 71:131–133.

Yang, C., Tana, Y. S., Harkema, J. R., and S. Z. Haslam. 2009. Differential effects of peripubertal exposure to perfluorooctanoic acid on mammary gland development in C57BL/6 and Balb/c mouse strains. *Reproductive Toxicology* 27:299–306.

Yang, L., Li, J., Lai, J., et al. 2016. Placental transfer of perfluoroalkyl substances and associations with thyroid hormones: Beijing prenatal exposure study. *Scientific Reports* 6:21699. doi: 10.1038/srep21699

Yang, Q., Xie, Y., and J. W. DePierre. 2000. Effects of peroxisome proliferators on the thymus and spleen of mice. *Clinical and Experimental Immunology* 122:219–226.

Yang, Q., Xie, Y., Ericksson, A. M., Nelson, B. D., and J. W. DePierre. 2001. Further evidence for the involvement of inhibition of cell proliferation and development in thymic and splenic atrophy induced by the peroxisome proliferator perfluoroctanoic acid in mice. *Biochemical Pharmacology* 62:1133–1140.

Yang, Q., Abedi-Valugerdi, M., Xie, Y., et al. 2002. Potent suppression of the adaptive immune response in mice upon dietary exposure to the potent peroxisome proliferator perfluorooctanoic acid. *International Immunopharmacology* 2:389–397.

Yang, Q., Xie, Y., Alexson, S. H. E., Nelson, B. D., and J. W. DePierre. 2002a. Involvement of the peroxisome proliferator-activated receptor alpha in the immunomodilation caused by peroxisome proliferators in mice. *Biochemical Pharmacology* 63:1893–1900.

Yang, Q., Abedi-Valugerdi, M., Xie, Y., et al. 2002b. Potent suppression of the adaptive immune response in mice upon dietary exposure to the potent peroxisome proliferator, perfluorooctanoic acid. *International Immunopharmacology* 2:389–397.

Yarwood, G., Kemball-Cook, S., Keinath, M., et al. 2007. High-resolution atmospheric modeling of fluorotelomer alcohols and perfluorocarboxylic acids in the North American troposphere. *Environmental Science and Technology* 41:5756–5762. doi: 10.1021/es0708971

Yeung, L. W. Y., Loi, E. I. H., Wong, V. Y. Y., et al. 2009. Biochemical responses and accumulation properties of long-chain perfluorinated compounds (PFOS/PFDA/PFOA) in juvenile chickens (*Gallus gallus*). *Archives of Environmental Contamination and Toxicology* 57:377–386.

Ylinen, M., Kojo, A., Hanhijarvi, H., and P. Peura. 1990. Disposition of perfluorooctanoic acid in the rat after single and subchronic administration. *Bulletin of Environmental Contamination and Toxicology* 44:46–53.

Yoo, H., Guruge, K. S., Yamanaka, N., et al. 2009. Depuration kinetics and tissue disposition of PFOA and PFOS in white leghorn chickens (*Gallus gallus*) administered by subcutaneous implantation. *Ecotoxicol. Environ. Saf.* 72(1):26–36.

Young, C. J., Furdui, V. I., Franklin, J., et al. 2007. Perfluorinated acids in Arctic snow: New evidence for atmospheric formation. *Environmental Science and Technology* 41:3455–3461.

York, R. G., Kennedy, G. L., Olsen, G. W., and J. L. Butenhoff. 2010. Male reproductive system parameters in a two-generation reproduction study of ammonium perfluorooctanoate in rats and human relevance. *Toxicology* 271:64–72.

Yu, N., Shi, W., Zhang, B., et al. 2013. Occurrence of perfluoroalkyl acids including perfluorooctane sulfonate isomers in Huai River Basin and Taihu Lake in Jiangsu Province, China. *Environmental Science and Technology* 47:710–717.

Yu, Q., Zhang, R., Deng, S., Huang, J., and G. Yu. 2009. Sorption of perfluorooctane sulfonate and perfluorooctanoate on activated carbons and resin, kinetic and isotherm study. *Water Research* 43:1150–1158.

Yu, W. G., Liu, W., Jin, Y. H., et al. 2009a. Prenatal and postnatal impact of perfluorooctane sulfonate (PFOS) on rat development: A cross-foster study on chemical burden and thyroid hormone system. *Environmental Science and Technology* 43:8416–8422.

Yu, W. G., Liu, W., and Y. H. Jin. 2009b. Effects of perfluorooctane sulfonate on rat thyroid hormone biosynthesis and metabolism. *Environmental Toxicology and Chemistry* 28:990–996.

Yu, W. G., Liu, W., Liu, L., and Y. H. Jin. 2011. Perfluorooctane sulfonate increased hepatic expression of OAPT2 and MRP2 in rats. *Archives of Toxicology* 85:613–621.

Zareitalabad, P., Siemens, J., Hamer, M., and W. Amelung. 2013. Perfluorooctanoic acid (PFOA) and perfluorooctanesulfonic acid (PFOS) in surface waters, sediments, soils and wastewater – A review on concentrations and distribution coefficients. *Chemosphere* 91:725–732.

Zeng, H. -C., Zhang, L., Li, Y.-Y., et al. 2011. Inflammation-like glial response in rat brain induced by prenatal PFOS exposure. *NeuroToxicology* 32:130–139.

Zeng, H. -X., He, Q. -Z., Li, Y. -Y., et al. 2014. Prenatal exposure to PFOS caused mitochondria-mediated apoptosis in heart of weaned rat. *Environmental Toxicology* 30:1082–1090.

Zhang L., Liu, J., Hu, J., et al. 2012. The inventory of sources, environmental releases and risk assessment for perfluorooctane sulfonate in China. *Environmental Pollution* 165:193–198.

Zhao, B., Li, L., Liu, J., et al. 2014. Exposure to perfluorooctane sulfonate in utero reduces testosterone production in rat fetal Leydig cells. *PLOS ONE* 9:e78888. doi: 10.1371/journal.pone.0078888

Zhao, L., Bian, J., Zhang, Y., Zhu, L., and Z. Liu. 2014. Comparison of the sorption behaviors and mechanisms of perfluorosulfonates and perfluorocarboxylic acids on three kinds of clay minerals. *Chemosphere* 114:51–58.

Zhao, W., Zitzow, J. D., Weaver, Y., et al. 2007. Organic anion transporting polypeptides contribute to the disposition of perfluoroalkyl acids in humans and rats. *Toxicological Sciences* 156(1):84–95.

Zhao, Y., Tan, Y. S., Haslam, S. Z., and C. Yang. 2010. Perfluorooctanoic acid effects on steroid hormone and growth factor levels mediate stimulation of peripubertal mammary gland development in C57BL/6 mice. *Toxicological Sciences* 115(1):214–224.

Zhao, Y., Tan, Y. S., Strynar, M. J., et al. 2012. Perfluorooctanoic acid effects on ovaries mediate its inhibition of peripubertal mammary gland development in Balb/c and C57BL/6 mice. *Reproductive Toxicology* 33:563–576.

Zhao, Z., Xie, Z., Moeller, A., et al. 2012. Distribution and long-range transport of polyfluoroalkyl substances in the Arctic, Atlantic Ocean and Antarctic coast. *Environmental Pollution* 170:71–77.

Zheng, L., Dong, G. H., Jin, Y. H., and Q. C. He. 2009. Immunotoxic changes associated with a 7-day oral exposure to perflurooctanesulfonate (PFOS) in adult male C57BL/6 mice. *Archives of Toxicology* 83(7):679–689.

Index

ATTRACTING SUSTAINABLE INVESTMENT

This book is a practitioner's guide to sustainable development, laying out strategies for attracting investment for communities and their partners.

It proposes an innovative Sustainable Development Proposition (SDP) decision-making tool based on a propositional calculus that can be used to analyse the sustainability of an infrastructure investment. It draws on environmental sustainability governance data analysis enabling investors to understand the economic indicators, income potential, return on investment, demand and legal compliance, as well as community and social benefits. Identified risks, issues and advantages are managed and monitored, and the SDP guidance can be applied to improve the prospects of the project in order to attract investment.

Sustainable Community Investment Indicators (SCIIs™) have been developed to assist with attracting investment and monitoring feedback on infrastructure projects, designed by the author for remote rural and indigenous communities – in response to current industry tools that are designed for urban environments. The book includes a broad range of real-world and hypothetical case studies in agricultural and indigenous areas in South America, Europe, Africa, Asia, Australia and the Pacific.

Taking a diverse economies approach, these industry tools can be adapted to allow for enterprise design with unique communities. This book provides sustainable development practitioners, including government agencies, financiers, developers, lawyers and engineers, with a positive, practical guide to addressing and overcoming global issues with local- and community-based solutions and funding options.

Saskia Vanderbent is a lawyer based on the Gold Coast, in sunny Queensland, Australia. She holds a multi-disciplinary PhD from the faculty of Engineering and a Juris Doctor in Law.

"Saskia's study of what works when investing in sustainable energy in remote and Indigenous communities will be a great boon to both those communities and potential investors. The text makes fascinating reading. Her conceptual tool, the Sustainable Development Proposition, makes application of that knowledge a little easier in practice."

Stephen Keim SC, *Barrister-at-Law,*
recipient of the Law Council of Australia's 2020 President's Award
and recipient of the Human Rights Medal, 2009,
by the Australian Human Rights Commission

"The tools in this book will ensure that your return on investment goes beyond a monetary return. You will have a sound basis to expect that your investments, and the partnerships that are formed with communities, are building capacity at the local level, and enhancing connectivity and opportunities with the broader economy in a way that is, dare I say, sustainable in perpetuity."

Craig Cowled, *Engineer, Researcher, Educator, Worimi man*

"Saskia Vanderbent has provided a comprehensive insight into the diverse energy challenges being confronted globally and how communities are moving to address them. This prescient perspective takes its currency in the present circumstances facing the world."

Allan Fife OAM, *Chief Investment Officer, Fife Capital*